G000151321

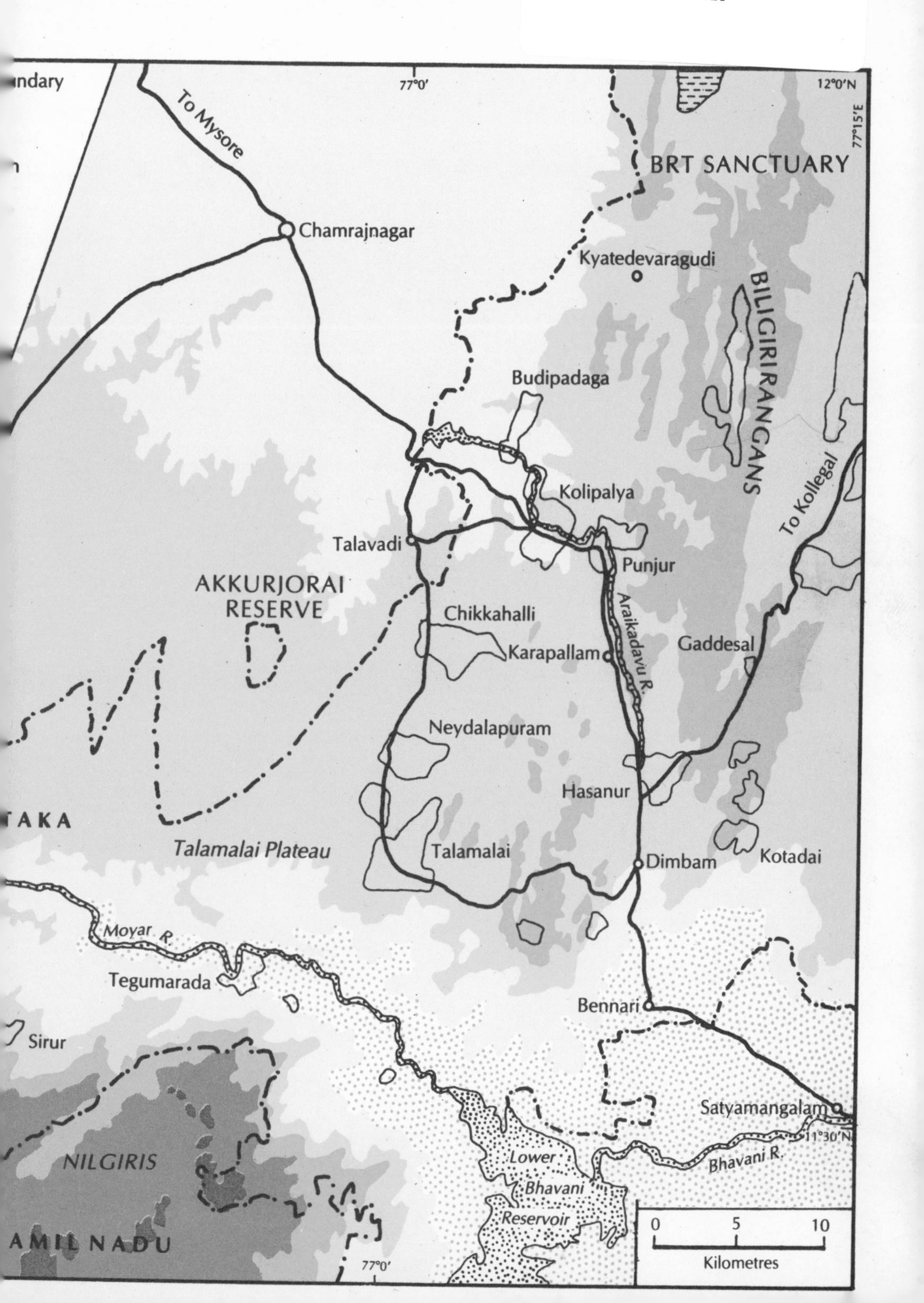
ndary
n
77°0'
12°0'N
77°15'E
To Mysore
BRT SANCTUARY
Chamrajnagar
Kyatedevaragudi
BILIGIRIRANGANS
Budipadaga
To Kollegal
Kolipalya
Talavadi
Punjur
AKKURJORAI
RESERVE
Araikadavu R.
Chikkahalli
Gaddesal
Karapallam
Neydalapuram
Hasanur
TAKA
Talamalai Plateau
Talamalai
Dimbam
Kotadai
Moyar R.
Tegumarada
Bennari
Sirur
Satyamangalam
11°30'N
Bhavani R.
NILGIRIS
Lower
Bhavani
Reservoir
AMIL NADU
77°0'
0
5
10
Kilometres

Elephant Days and Nights

Elephant Days and Nights

Ten Years with the Indian Elephant

RAMAN SUKUMAR

Foreword by George B. Schaller

Delhi
Oxford University Press
Bombay Calcutta Madras
1994

Oxford University Press, Walton Street, Oxford OX2 6DP
Oxford New York Toronto
Delhi Bombay Calcutta Madras Karachi
Kuala Lumpur Singapore Hong Kong Tokyo
Nairobi Dar es Salaam Cape Town
Melbourne Auckland Madrid
and associates in
Berlin Ibadan

ISBN 0 19 563348 2

Typeset by Taj Services Limited, Noida 201301
Printed at Rajkamal Electric Press, New Delhi 110033
and published by Neil O'Brien, Oxford University Press
YMCA Library Building, Jai Singh Road, New Delhi 110001

To the memory of
my mother

Contents

List of Illustrations

Foreword

Like others concerned with conservation, I have felt a terrible sadness and anger at the destruction of elephants for their ivory. And like most others, my mind was on African elephants. Then I read a superb monograph entitled *The Asian Elephant, Ecology and Management* by Raman Sukumar, published in 1989, in which he documents how the spread of agriculture, logging, and poaching has reduced the Asian elephant to a mere forty to fifty thousand individuals in small, scattered populations extending from India to China. The monograph not only focused my attention on the plight of Asian elephants, it also rekindled my interest in India where I had conducted wildlife research over two decades before. Not long afterward, I had an opportunity to return to India, indeed to the same forests in the south where Sukumar had studied and was still studying elephants. There, for the first time, I observed Indian elephants at leisure. One afternoon in Nagarahole National Park, biologist Ullas Karanth and I sat quietly by a pond. Screened by tufts of grass, we were submerged in the scene. Several jungle fowl scratched beneath clumps of bamboo, a side-striped mongoose hurried by, and four wild pigs rooted and wallowed near the pond's edge close to the pugmarks of a tigress. Behind us a peacock gave its strident call. Suddenly, six elephants silently emerged from a stand of teak and boldly entered the pond where, belly-deep, they drank, sprayed themselves, and sloshed around, enjoying the cool water after a hot and dusty day. This elephant family, bound by intimate ties of kinship, comprised three females, a sub-adult male, and two youngsters, one so small that it had to snorkel rather than bathe. We were part of this antediluvian spectacle, a wilderness experience as exciting as any that Africa has to offer. On detecting our scent, the elephants became fretful and, led by the matriarch, filed past us into the forest. I exult in an experience such as this but there is a tinge of sadness as well because I realize that elephants, like so many other species, have only a tenuous future.

Several days later, a few miles south of Nagarahole, I joined Sukumar in the Mudumalai Sanctuary. During one of our outings we watched a herd of elephants glide past on one side of a ravine, wholly engrossed in their own activities, while we stood quietly on the other side. A fire had swept through this *Anogeissus* forest leaving the ground bare and black. A flush of green shimmered in the late sun and the white tree trunks gleamed like ivory. Although elephants add a transcendent quality to any scene, this was not a wilderness. In the distance I could hear cars on a road. And later, in a nearby river, a domestic elephant bull reclined at ease while a man vigorously scrubbed his bristly back. People washed clothes along the river bank and two tiny elephant calves splashed like puppies in the shallows. In this part of India, domesticated elephant cows are released into the forests to be mated by wild bulls. There was a strange conjunction between wilderness and civilization in these elephants. One moment we saw them as living monuments to the past and symbols of the vanishing forests. The next they evoked visions of the pomp of kings and emperors, and of docile beasts of burden hauling logs out of forests, ironically assisting in the destruction of their home. They seem lost between two worlds.

During this visit to Mudumalai, Sukumar told me that he was just finishing a popular book on his decade-long elephant research. *Elephant Days and Nights* is his personal covenant to help the species. By 'studying them, photographing them, chasing them and being chased by them', he became addicted to elephants and the 'junglee life', as he calls it. Elephants past and present are intricately interwoven in India's culture but this human dimension to their existence does not resolve the problems they cause. Confined to small forest tracts they venture forth to plunder fields, push over coconut trees, and on occasion kill an unwary villager. In turn, they are shot in retaliation or poached for their ivory. We know how to protect elephants, but not how to manage them adequately. Management requires knowledge. By studying how elephants raid crops and which individuals cause the most damage, Sukumar has been able to suggest innovative strategies for reducing conflict. In this, he has made a major contribution to helping elephants endure as icons of India's culture.

Elephant Days and Nights is filled with fascinating facts. One domestic elephant cow lived at least seventy-nine years. About five per cent of the elephant bulls in southern India are genetically tuskless and in Assam the figure is nearly fifty per cent, certainly a pre-adaptation against ivory poachers. And 'what do you do when faced with an angry elephant?' Good research demands both a passion to understand and emotional involvement. In the pages of this book we meet Kali, Tara, Meenakshi, Vinay, and other elephants, known individuals based on years of contact, each a complex being with its own aspirations and problems based on family ties and traditions. Present in India for at least four million years, since the mid-Pliocene, elephants have constantly responded to and evolved in their ever-changing world, and this book's descriptions of their life are graceful, evocative, and acute.

Books such as this, both passionate and pragmatic, forge a mental link between our fate and that of other species. They remind us of today's most urgent issue: to conserve the diversity of life. That the elephant has so far survived in crowded India is in itself a statement of the nation's vision and of the social and spiritual value that people place on the elephant. Ganesha, the elephant-headed god in Hindu mythology, represents the great power that is inherent in all of us. By his dedication to elephants, Sukumar is using this power to create a wider awareness of and deeper empathy for one of this planet's most magnificent creatures.

GEORGE B. SCHALLER
WILDLIFE CONSERVATION INTERNATIONAL

Preface

Even today, after more than twelve years of study and observing thousands of elephants, I cannot help but just stop and look whenever I see one. If the tiger is the spirit of the jungle, secretive and elusive, the elephant is its body, large, majestic, making its presence felt with authority. Yet, an elephant's life seems so unhurried; it is almost like watching a movie in slow motion. To even begin to understand this fascinating animal one has to spend several years in the field. Every day, every month, every year, one learns some new facet of its character. Even several years may not be sufficient to understand details of its population dynamics; this may require several decades of data gathering. An elephant, after all, lives out the proverbial life span of three score and ten years.

The Asian elephant today faces an uncertain future. Its plight is not as well appreciated as that of its African cousin. The main reason for this is that, unlike the African elephant, the decline of the Asian elephant has historically been a very gradual process. Virtually unnoticed by the international community of conservationists, the Asian elephant has slipped to an estimated worldwide population of less than 65,000 individuals, only a tenth that of its African relative. This calls for an urgent response, not merely emotional but also pragmatic, from individuals, societies and nations. The next decade could decide the ultimate fate of a creature whose relationship to the natural world and the human spirit has and will remain unrivalled.

My study of elephants in southern India began in 1980 and continues to the present. In this book I have tried as far as possible to maintain a chronological narrative of my study, with each chapter emphasizing a particular aspect of the elephant's life. This has necessitated some interpolation of observations that do not strictly follow a chronological sequence. The narrative interweaves the biology of the elephant with my personal field observations. Wherever appropriate I have inserted edited versions of my notes to give the reader a feel for the action in the field. Often, the photographic record I maintained helped me to capture details I had missed in the field.The camera was thus an invaluable extension of my note-book. I have, therefore, chosen photographs not necessarily for their pictorial quality but in order to illustrate the events described in this book.

The first seven chapters give an account of the early part of my study from 1980 until about 1983 with some insertions of observations made in later years. The eighth chapter carries the story forward up to 1990. Chapter 9 goes back 4,000 years and traces the historical association between elephants and people, summarizes the present status of the elephant and makes suggestions for its conservation. A postscript brings the story up to the present.

In a study spanning over a decade it is but natural that I have received the help and support of a large number of individuals and institutions. Instead of listing each one of them, something I have done in detail in my earlier publications, I have chosen to express my appreciation by making them a part of the story. It is inevitable that only those who played a relatively major role in my work could be fitted into the narrative. I do sincerely thank all those who have played a role, whether major or minor, in my study, with the fervent plea to those not explicitly mentioned here that they do not bear this in their memories anywhere as long as an elephant would!

CHAPTER 1

The Problem

Chance remarks and luck have often played an important role in my life, and the way in which I came to study gorillas was no exception.

George B. Schaller in *The Year of the Gorilla* (1964)

Biligiri seemed unsure of what to do with himself. The entire afternoon he had been hesitant, shuffling up and down the welltrodden path that ran through the jungle to the pond. He listened to the sounds of music and laughter, of a child wailing, the squeak of a wheel cranking up a bucket of water from a well, of wet clothes being beaten against a stone, but he did not really take them in. Not today. They seemed rather jumbled. He was undergoing a strange transformation, a transformation which was perfectly normal for his age, though he did not know it. His entire body was alight, there was a dull ache in his temples, his blood seemed to surge with a new urgency. For the first time he was in musth. And he was confused, like other sixteen year olds. Biligiri was an adolescent wild male Asian elephant. He was at Kyatedevaragudi in the Biligirirangan hills of southern India. The date was 19 January 1983.

He plunged into the water, then heaved himself up only to plunge back in again. He stood up and thrashed the water with his trunk, then lay on his left side, turned onto his right side and blew the water from his trunk like a fountain. He then jerked his head down, slapping the surface of the water with his tusks and then suddenly, in a moment of frenzy, immersed his head beneath the surface to dig his tusks into the pond bottom. That seemed to give him some relief. Never since his childhood days had he given himself to such abandon, but then he had never been in musth before. Biligiri was totally oblivious to the large crowd watching him from the edge of the pond only a short distance away.

The sun was a mellow red ball at the edge of the vast plains to the west when Biligiri climbed out of the pond and slowly made his way back through the jungle, his nerves sufficiently soothed for the moment. The final break with his family had come some time ago. When the others had moved down from the hills to the southern valley he had stayed back. He now had to chart out his own course and make his own decisions. He would wander alone, associating with other bulls when he had the chance or joining a herd if there was a cow he found sufficiently interesting.

Would he grow to be a gentle soul, not capable of harming a child, or a rogue, trampling paddy fields and people alike? Would he live to three score and ten years, or would he end his life prematurely in a pool of blood with his tusks cruelly axed from his skull? Only time would tell.

I was one of those watching Biligiri that evening. Also watching him were some thirty biology students who had come there on a nature camp. What better lesson could there have been for them, or for that matter for me, than Biligiri in this mood?

By now I had been watching him and others of his kind for over two years. I was studying them, photographing them, chasing them and being chased by them. I have seen my share of elephants courting, mating, giving birth, parenting, playing, fighting, rolling in slush, enjoying life, and dying. Elephants can be funny. Elephants can be frightening. They can be heart-warming and their study can be addicting. I must confess that I am a victim of this last mentioned affliction. I have often been asked how I came to be involved with elephants and what I really do with them. This book is my attempt to tell the story, not only of my own experiences with elephants, but also of their lives and the problems they face today.

My mother was firmly convinced that all the *junglee* movies that my father had taken me to when I was young were responsible for my 'strange' pursuit of elephants. Such a thing was just not the tradition in Indian families! Her observation would have been welcomed by those who strongly believe that environmental influences rather than genes shape an individual's behaviour. There was, however, no hint during my younger days of what I was eventually going to do. I was the usual city-bred child interested in games and various hobbies. I was very fond of books, an interest I attribute to my maternal grandfather who was a bookseller. One of my chief boyhood delights was to regularly raid his shelves. My paternal grandmother seems to have recognized a streak of the naturalist in me quite early on. I recall that for some reason she used to call me a *vanavasi* much before I ever went to a forest.

It was only when I reached high school that my eyes opened to the profound beauty of living creatures. Those were the heady days of space exploration and man's first landing on the moon. The mysteries of outer space beckoned more excitingly than the realities on Planet Earth. At that time the world was also waking up to the crisis facing our environment caused by pollution and the reckless exploitation of nature. The modern conservation movement was gathering momentum. Even then I was convinced that humans were heading for trouble unless they changed course towards a more prudent, efficient and equitable consumption of the planet's resources.

I had a passion for science and wanted to work in conservation. Nothing seemed more ideal than becoming a biologist specializing in ecology. With this goal in mind I pursued botany at Loyola College and later at Vivekananda College in Madras. I was fortunate to have teachers who encouraged creative pursuits outside the framework of the regular syllabus. I was also fortunate to be in Madras, the largest city in southern India, which is almost unique in having a national park, the Guindy National Park with its semi-natural forest, large populations of spotted deer and blackbuck, and a rich fare of birds, all within the city.

I spent most of my weekends at Guindy Park in the company of R. Selvakumar, who was pursuing his studies in zoology, and who was to accompany me on many future trips to the jungle. Selvam was the quintessential naturalist lost in a world of feathered and furry friends. He had an amazing ability to identify and describe anything from an ant to a whale and I must say that my work was made all the more easy and pleasant because of him.

I still did not have the faintest idea of getting involved with elephants although a passion for these pachyderms may have been latent in me. Baby Elephant Walk, that delightful song from the movie *Hatari* was and still is one of my favourites. I remember this was the first item I played when I compered for the local radio station.

My first sighting of a wild elephant was not spectacular; in fact, I doubt whether I saw an elephant at all. Our class was on a botanical tour during September 1976. We were in Mudumalai, a sanctuary in Tamil Nadu state in southern India which is well-known for its large elephant population. I was talking to P. Padmanaban, the Warden of Mudumalai, who was to later become the state's Chief Wildlife Warden and who was to give a boost to wildlife research, when a rickety car rolled in and a familiar figure leaned out and boomed, 'Hey, boy. What are you doing here?' It was the flamboyant Siddharth Buch, a naturalist-photographer and one of my early gurus in wildlife pursuits, who was visiting Mudumalai along with his brother-in-law. They had driven the 600 kilometres from Madras in the latter's old car, which would have boasted a top speed of 40 kilometres per hour on the highway. There was probably no other vehicle less suitable for going about in elephant country; at a mere swish of a jumbo's trunk its flimsy body would have fallen apart, and there were plenty of jumbos in Mudumalai. Yet here were these two brave gentlemen all the same.

Siddharth was delighted to see me. He enquired whether I had seen any elephants. I explained to him that we had only just then arrived. He immediately asked me to hop into their old jalopy as they were going for a drive along the Moyar river. Whatever my apprehensions about the car, my eagerness to see some wildlife made me obey instantly.

We had hardly gone a kilometre when Siddharth stopped the car, leaned out, pointed towards a hillock and said, 'There is an elephant moving up the slope. He must have just come up from the river and crossed the road.' I strained my eyes in that direction but could not see any elephant. 'Can't you see that black shape moving through the bushes? That's the back of an elephant', he continued. It was mid-day and the jungle was a patchwork of light and shadow. I guess I did see something darker than a shadow a hundred metres away but I could not see it moving. It may have been just a rock, but I nodded dutifully, meaning that I could indeed see a dark shape. 'There, my boy! I have shown you your first wild elephant!', announced Siddharth triumphantly.

Whatever the true identity of that object, with the benefit of hindsight I must caution readers unaccustomed to the capricious ways of wild elephants that it is always better to mistake a rock for an elephant than an elephant for a rock.

In August 1979 I joined the doctoral programme in ecology at the Indian Institute of Science in Bangalore. Madhav Gadgil had established the country's first academic programme in modern ecology. When the time came to select a research topic for my thesis, Madhav suggested some problems, among them a study of elephant–human conflict. My ears pricked up when he mentioned elephants. I had wanted to work on large mammals and what was better than studying elephants, although I must confess that such a thought had never occurred to me earlier. Madhav explained that in many areas wild elephants entered cultivation to feed on crops and sometimes killed people, while people made their own impact on the elephant's habitat and killed them in defence of their crops or in order to steal their tusks. This two-way interaction of elephants and man had never been studied in detail before in either the Asian or the African elephant. In that respect it would be a novel study; it would also be important from a conservation angle because the satisfactory resolution of this conflict was crucial to the long-term survival of elephants.

Around this time, in early 1980, the Asian Elephant Specialist Group (AESG) of the International Union for Conservation of Nature and Natural Resources (IUCN, now known as the World Conservation Union) was in the process of launching its first set of projects. J. C. Daniel, the then Chairman of the AESG, wrote to Madhav indicating that funds might be available to the group for a number of projects in India. The funds actually came from World Wildlife Fund–International (WWF, now called the World Wide Fund for Nature). I developed a proposal to study elephant–human conflict in southern India with the aim of developing management plans for elephant conservation and forwarded it for their consideration. M. A. Partha Sarathy, who was co-ordinating the Southern Indian Task Force of the AESG, also enthusiastically supported our project.

The previous year the Asian Elephant Specialist Group had met at our institute in Bangalore. Prior to the meeting a census of elephants had been conducted by the forest departments of the southern Indian states. This was the first time that a census of this kind covering the entire elephant range in the south had been attempted. When the figures were presented at the meeting there was considerable disagreement over the tally of 1268 elephants for the Satyamangalam and Erode Forest Divisions in the Eastern Ghats of Tamil Nadu state. Many members did not believe that this region could hold so many elephants; after all a figure of only 446 was presented for Mudumalai Sanctuary which was the famous elephant area of the state. What they did not clearly realise then was that the former region was much larger in area, over 2467 square kilometres, and hence actually had a much lower density of elephants according to the census than did Mudumalai which was only 321 square kilometres in extent. Anyhow, the Eastern Ghats region had many enclaves of cultivation where conflict between elephants and people seemed common and, therefore, was a potential area for my study.

Once I got through with my courses in May 1980 Madhav suggested that I do a preliminary survey of the Eastern Ghats to look at its elephant population and the areas of conflict. We had not yet received funding for the work but the Salim Ali Conservation Fund came to the rescue with a small grant towards travel expenses. Selvam was just beginning his research on the ecology of the gaur, the largest of all the wild Bovidae, in Mudumalai. He was also interested in a quick survey of gaur populations elsewhere before getting down to detailed work and so we decided to do a joint trip to the Eastern Ghats.

The afternoon of 27 May we rolled out of Bangalore in an old jeep whose fitness for moving about in elephant jungle was as questionable as Siddharth's brother-in-law's car. We had gone only a short distance out of Bangalore when the battery slid back from its stand and destroyed the voltage regulator located behind it. With a lot of spluttering and splattering the jeep continued on gamely until we reached the next town Kanakapura, fifty kilometres away, where an electrician's skill with a soldering rod saved us the ignominy of going back to Bangalore for a new regulator on the very first day. We could not get lodgings for the night and so stretched out our sheets inside the compound of the local forest office.

Kanakapura was not too long ago a walled city virtually surrounded by jungle. During the last two years of the reign of Tippu Sultan (1797–9), tigers were supposed to have claimed forty-eight lives inside this town! This, however, was 1980 and, although the jungle was not too far away, we were not disturbed by any tigers during the night.

The next seven weeks we explored the Eastern Ghats, visiting villages, enquiring about crop depredation, looking for signs of elephant and gaur in the jungle, trying to locate them, and checking out whether the habitat was contiguous or fragmented. We began with the Kanakapura Range, where cultivation had made considerable inroads into the degraded thorn forest. Hunsanhalli, one of the villages there had been raided by a large herd of elephants. It seemed that the village cultivation had spread in such a manner that it lay in the elephants' path when going for water to a small reservoir nearby. We did not see the elephants but the signs they had left behind in the fields and the adjacent forests were obvious.

Our jeep had by now begun to act up again. It would run a hundred metres or so and then stop. To get it going again would take quite a bit of persuasion. Finally, when going up a small hill in the jungle, it came to a halt. No amount of persuasion would make it respond. Pushing it up the hill was out of the question. So we tried to turn it around and start it by rolling it down the slope. We did turn it around only to realise too late that neither of us was in the driving seat to stop it! To our horror the jeep rolled down the side of the hill into a bamboo clump which mercifully halted its progress into the ravine below.

We sat by the roadside wondering what to do next. The jeep had first to be started before we could attempt anything else. In luck, a lorry with some two dozen men soon came by. We thought that with so many men the jeep could be pushed up to the road, but the lorry driver was made of sterner stuff. He simply got in behind the steering wheel and the jeep ignition responded at the first try! I have always wondered whether the shake up had made something fall into place or whether a healthy respect for burly lorry drivers made the jeep obey instantly. With the aid of the four-wheel drive and a bit of pushing the jeep was up on the road in a couple of minutes and so we continued on our trip.

South of Kanakapura the dry deciduous forest along the banks of the Cauvery river is virtually uninterrupted for nearly 100 kilometres between the Sivasamudram Falls in the west to the Stanley Reservoir at Mettur in the south-east. At Sangam, the confluence of the Arkavati and the Cauvery, a bull elephant had been regularly crossing the Cauvery to enter a coconut farm here. The farmer had connected a tape recorder to a loudspeaker through which a tiger's roar and a jumble of noises were broadcast. This met with some success.

Five kilometres downstream at Mekkadatu ('goat's leap') the

waters of the Cauvery flow rapidly through a rocky, narrow gorge. The entire region is hilly and animals such as elephants have access to the perennial waters only at certain places. There seemed to be few gaur north of the Cauvery, though they were relatively common to the south.

From the Cauvery we drove on to the Biligirirangan hills. As we approached the hills from Chamarajanagar town they seemed to shimmer a blue colour (as many other hill ranges do) and not the white their name implied. Much later I learnt how they had been named.

We saw our first wild elephants here. We had just entered the open-canopied scrub jungle of the plains before the climb up the hills, when around a curve we suddenly chanced upon a herd silently crossing the road in single file, their backs covered with red soil. Selvam remarked that it was a scene reminiscent of the African bush. More sightings were to follow.

We stayed in the bungalow at Kyatedevaragudi, overlooking the Mysore plains to the west. At the pond just behind the bungalow we saw herds of elephants and lone bulls. It was now June and the monsoon had set in. The forest was a lush green, the trees had flushed new leaves, the ground was a dense tangle of ferns and grasses and the motorable tracks were overgrown. Trees had fallen across the roads but Masti, our old Sholaga tracker, disposed of them in no time with his billhook. There were certainly plenty of elephants in these hills at this time of the year. Although we had no objective way of censusing them during a short visit like this a guess that there were some 500 elephants did not seem unreasonable.

Selvam was also not disappointed; the moist deciduous forest was good gaur country. We went up to Attikan, a century-old coffee estate and stayed in the picturesque bungalow there. From there we walked up to the summit of Kattari Betta, the highest peak, to absorb the panorama of rolling grasslands and stunted montane forests nestled amidst the hollows. There were few signs of elephants here.

Our next destination was the Satyamangalam Forest Division to the south. We descended the hills to Punjur, where we heard more tales of crop depredation. From Punjur we drove through the Araikadavu valley, crossing over into Tamil Nadu state. This valley was to later become a happy 'hunting' ground for elephants. About a kilometre before Hasanur, the first village on the Tamil Nadu side, we spotted a pink, tile-roofed building, seemingly standing in isolation at a distance on a hillock jutting above the low forest canopy. I did not realise then that this was soon to become my home for the next two and a half years, so we drove on past Hasanur thinking that the forest bungalow where we were scheduled to stay was further away at Dimbam.

In the jungle midway to Dimbam we saw a man apparently sleeping by the roadside. We thought this a bit strange and that he was probably drunk. We continued to Dimbam where we learnt that the forest bungalow was at Hasanur. On our way back we noticed some men near the man we had seen earlier and stopped to enquire about him. The men were digging his grave. He had been killed by an elephant the previous evening. No one knew his identity; he seemed to be a wanderer. There were large footprints of a single elephant nearby, obviously an adult bull. In our naïvety we were hardly expecting this but this was the stark reality one has to keep in mind when talking about elephant conservation.

We stayed at Hasanur for a few days, going around the forests and visiting the villages. Hasanur and the villages to the west on the Talamalai plateau seemed to be targets of regular elephant depredation during the crop season from September to December. We drove down to the town of Satyamangalam, situated on the banks of the Bhavani river, and attempted to go up the hills on the other side to Kadambur in the east. The road up the ghats was a medley of large stones and rocks. After two or three kilometres we gave up and turned back. Our jeep having behaved exceptionally well during the previous few days, we did not want it to be pushed beyond its limits.

Two years later when I finally managed to get to Kadambur with a forest officer I learnt how a vehicle should be manœuvred on this road. We met a bus proceeding along the road with the radiator grill removed and two men sitting in the space it vacated. At first I thought that they were sitting there because the bus was crowded inside; after all it was not uncommon for people to sit on the roof of a bus. I was mistaken. The two men were in that position so that they could quickly hop off and clear away any boulders that may obstruct the bus. They had plenty of work to do. Since neither Selvam nor I had any experience in sitting on the front bumper of a moving vehicle, we probably could not have taken our jeep up the ghat road in this fashion.

We branched off the Hasanur road and proceeded northwards to Kollegal. Although there was potentially good elephant country along the way, there were also plantations of eucalypts. A large settlement for refugees from Tibet had considerably reduced the habitat.

That the Kadambur road had taken its toll on our jeep we soon discovered. We halted at a village and upon starting again heard a loud thud; a ring in the self-starter had come loose and fallen into the gear box. Fortunately, the villagers gave us a push and we drove without stopping until we reached a repair shop at Kollegal.

From Kollegal we headed east to Madeshwaramalai and then south to the Bargur hills. These hill ranges are extensive and though there were certainly many elephants here it was very difficult to actually see them. We were driving along a narrow disused track in the Bargur hills, in not much better shape than the one to Kadambur, barely squeezing our way through the dense undergrowth of thorny *Lantana* shrub when, as we reached the summit, we noticed fresh elephant dung on the track. Jumping out from the jeep we went to examine the warm droppings when a noise, not unlike that of battle tanks, came crashing through the bushes right next to us, sending us diving back into the jeep. A herd of elephants was stampeding—fortunately in a direction away from us! They could hardly have been ten metres away from us and yet we had not even glimpsed their backs, such was the growth of *Lantana*. The herd, which must have been large since there were dung piles of various sizes, had walked along the track for over a kilometre before meeting us.

The Hosur-Dharmapuri Ranges which are contiguous with the Kanakapura Range to its east were our final destination. The dry deciduous forests here were considerably degraded due to the pressure from human settlement. We heard of a limping rogue tusker haunting the villages. Many of the bulls were being killed for ivory or when they came into the crop fields.

Our woes with the jeep were not yet over, however. The fuel pump refused to do its job whenever it got heated, which was quite often during the heat of the Indian summer. We had wrapped a moist cloth around the pump. On one occasion when the cloth became dry it caught fire. When we opened the bonnet and saw the smoke billowing out I first thought that everything was going to blow up. Fortunately, Selvam had the presence of mind to pick up a stick and prise the flaming cloth away.

After returning to Bangalore in mid-July I prepared a report on the status of elephants and conservation problems in the Eastern Ghats. Based on our actual sightings, the information from villages on raiding groups, and indirect signs and inspired guesses, I made density estimates for different regions. The final tally was about 2000 elephants for a forested area of some 7000 square kilometres, a figure not very different from the forest department's own census.

The Asian Elephant Specialist Group met in August that year at Colombo in Sri Lanka. My colleague Vijayakumaran Nair, who had earlier surveyed other elephant populations and I attended the meeting. We presented the distributional picture for southern India. Everyone was happy at the 'discovery' of these 2000 elephants in the Eastern Ghats. I also put up my proposal for the detailed study of elephant ecology focusing on elephant–human interactions. Following the meeting we were taken to Yala National Park in the south where we were treated to a rich wildlife spectacle including plenty of elephants. The Sri Lankan national parks and wildlife reserves, headed by Lyn de Alwis, were some of the finest in Asia, shortly before the tragic civil war threw them into disarray. At Yala, as in other parts of the country, the male elephants are mostly tuskless. Whether this was always the case and the few tuskers had inherited their genes from captive tuskers imported from southern India by the island's rulers or whether the tuskers had been wiped out by selective hunting and capture is still a mystery. We saw just one tusker and were charged most impressively by a tuskless one.

Upon returning to India I decided to base myself at Hasanur in the Satyamangalam Forest Division of Tamil Nadu. From here I could conveniently explore the entire Biligirirangan range, the Talamalai plateau and even go down to the Moyar river valley when necessary. The forest bungalow had a spare room I could use and came with an added bonus. Subramaniam, the caretaker of the bungalow, was an excellent cook. Without having to worry about daily chores I could devote more time to my work.

In October I went to Madras to meet T. Achaya, the Chief Conservator of Forests in Tamil Nadu, to obtain permission for my proposed elephant study. No sooner had I completed my first sentence than he picked up the phone and instructed the concerned officials to make arrangements for my stay and work at Hasanur. I have never experienced a quicker response from a government official. My study was off to a flying start. Soon I was talking to K. Venkatakrishnan, the Conservator of Forests at Coimbatore. He too was enthusiastic about my proposed study. From Coimbatore it was onwards to Satyamangalam to meet C. R. Thirumurthy, the District Forest Officer, and to finalize the logistics for beginning the study. At Hasanur I was greeted by Premnath, the slim Ranger, who provided the space I needed in the forest bungalow.

One night I was standing near the police check-post at Hasanur when an excited driver jumped out of his lorry. A few minutes earlier he had spotted the largest elephant he had seen in his life by the roadside some two kilometres before the check-post on the way from Dimbam. The elephant apparently had only a single tusk on its left side. The driver loudly wondered what would have happened if the bull had decided to put his tusk through his lorry.

I was to hear and see much more of the exploits of this bull in the months to come. It was one of the most notorious and aggressive elephants in the area, the *bête noire* of Hasanur. Later I discovered that its right tusk was not absent but broken off at the base. I first named the elephant Ganesh, after the elephant-headed god of the Hindu pantheon pictured with a broken right tusk, but later changed his name to Vinay, after Vinayaka, another epithet of the elephant god depicted in early mythology as a creator of obstacles . . . but more of this later.

Over the next few days I received complaints from the farmers of Hasanur about the depredations of this bull. I saw trampled paddy fields, half-eaten millet plants and pushed-over coconut trees. It was only a foretaste of the damage I was to later witness during the study.

From Hasanur I proceeded to Mudumalai, which was to be a minor study area. Selvam had already commenced his study of the gaur here. The first afternoon we saw another large bull elephant with which we both later became well acquainted. Not being still fully aware of what elephants are capable of, we rather foolishly spent some three hours very close to the bull. In luck, it turned out to be one of the most gentle and tolerant bulls we have ever encountered. I exposed over a hundred frames of film on this elephant, which was feeding on the bark of a young teak tree. The bull's tusks were long and spread outwards. We called it Divergent Tusks. It must have been one of the most photographed among Asian elephant bulls in those days. Years later I instantly recognized its picture in a Los Angeles home.

When Selvam and I met V. M. Narasimhan, who had recently taken over as Warden of Mudumalai, he guffawed loudly, his eyes almost bulging out, 'You *bachas*, one wants to study gaur and the other elephants! Make sure you know which is the front end of an elephant! And keep away from its trunk!'

Narasimhan was a tough, six-foot, ex-army man of a breed that has now all but vanished. Beneath his tough exterior he was a warm-hearted person. He loved to pull one's leg, but if at the end of an hour you had survived he would grant you anything you wanted. Nothing pleased him more than to see a young person taking to field research. Selvam and I had survived his company since our Madras days when he had by coincidence been the Warden of the Guindy Park.

When I returned to Bangalore I found a letter from J. C. Daniel stating that WWF–International had confirmed support for the Indian projects on Asian elephant conservation. Three major projects were to be launched, two involving surveys of elephant habitats in north-eastern and central India, while the third was to be my study of elephant–human conflicts in southern India. I was assured of funds toward hiring a field tracker and expenses for running a vehicle.

I still needed funds to buy a vehicle. Fortunately my parents came up with the money. In December I located a nineteen year old jeep at Madras and spent the first month of the new year in overhauling it for jungle life. In this I was assisted by Jagannatha Rao, an engineer turned naturalist, whose tips on the finer aspects of vehicle maintenance proved invaluable in keeping it on the road through sun, rain, dust, stones and elephants. I was not keen to repeat my earlier experiences with a jeep during the preliminary survey.

On 27 February 1981 I drove off to Hasanur, accompanied by Jagannatha Rao who came along to help me set up camp, to begin the intensive phase of the field work. As we approached Hasanur late in the evening we were greeted by a bull with only one visible tusk, standing near a bridge. He merely turned away from us upon seeing our lights and went into the dense *Lantana* bush.

CHAPTER 2

The Land and the Elephants

Listen! This is the great god of *Dodda Sampige*,
My Lord with a thousand tresses;
Behold how beautifully he comes!

A song of the Sholagas.

Did this beautiful hill praise Andiran who wields a sharp sword and wears a wreath of *carupunnai*? How is it this hilly forest is full of tuskers?

A poem describing the land of the Cheras. Tamil Sangam Literature (1st–5th century AD).

Six kilometres to the south of Hasaṇur, on a clear day the bungalow at Dimbam perched at the edge of the plateau commands a panoramic view of the Coimbatore plains. One morning, a week after I reached Hasanur, I stood at this point looking down at a vast carpet of white cloud which had settled in the plains overnight. It was more like a scene from a plane flying at 10,000 metres than one from a hill only 1000 up. As the sun rose on the horizon the vapourized cloud wafted upwards, reducing visibility to a few metres. It took nearly two hours for the fog to clear.

The road from Dimbam to Satyamangalam, twisting its way down through twenty-seven hair-pin bends along the escarpment, gradually came into view. The thorn jungle in the plains, still green in patches, stretched out for some distance before giving way to the vast expanse of agricultural fields. The multicoloured *gopuram* of the temple at Bennari could be seen amidst the brown jungle, a kilometre inside its distinct boundary with the fields. A bus strained up and around the treacherous twentieth hair-pin bend, the whine of its engine dying momentarily as the driver shifted to first gear. Then I heard the unmistakable sound of an elephant trumpeting in the jungle below.

Setty, the Irula tracker I had hired a couple of days earlier, located the animals within half a minute. Scanning through my binoculars I picked up two black dots, one larger than the other, in an open patch. The trumpeting continued for some time; perhaps they were contact calls with other members of a herd. Over the next two years I found that in early March or thereabouts the elephants invariably left the thorn jungles in the plains for higher altitude forests.

During March and April 1981 I explored the main study area. Since the area was part of a much larger forested tract, I also had to define the boundaries of my study population. I wished to cover an area that was large enough to encompass the home ranges of several elephant herds but was also manageable logistically. Eventually I defined an area of about a thousand square kilometres which included an entire spectrum of vegetation types from wet evergreen forest to dry scrub jungle.

The northern limit of my study area was an artificial line running across the settlement at Biligiri Rangaswamy Temple Betta (or simply BRT Betta); to the south, the boundary was in part the natural one between jungle and cultivation in the Coimbatore plains and in part the Moyar river separating the Biligirirangans from the Nilgiris. The highway connecting the town of Kollegal to Satyamangalam was a convenient eastern limit, while to the west the boundary between forest and agriculture on the Talamalai plateau was sharply defined. To the south-west the elephant habitat is contiguous along the Gazalhatti pass (originally Gajapatti or 'abode of elephants') in the Moyar valley into the Bandipur National Park and Mudumalai Sanctuary. These latter areas remained minor study sites for the present.

The main study area is geographically known as the Biligirirangan hills, although this name is today mistakenly restricted to the northern portion falling within the state of Karnataka. *Ranga* is the presiding deity of the ancient temple atop the precipitous mass of granite along the western face of the hills. The archaeological literature does not give a firm date to its construction, but the Dravidian architectural style suggests that the temple may have originally been constructed sometime between the tenth and twelfth centuries. *Biligiri* literally means white mountain, a rather unusual combination of a Kannada word (*bili* = white) and a Sanskrit word (*giri* = mountain). For this reason, the legendary G.P. Sanderson, who captured elephants here during the nineteenth century, believed that the correct term is *Bilikul-ranga* or *Biliga-ranga* as pronounced by the local people (*kul* meaning stone in Kannada) and referring to the deity atop the white rock.

The two central ranges of hills running parallel in a north–south direction for eighty kilometres, from Kollegal to Satyamangalam, form the backbone of the Biligirirangans. From Kattari Betta, at an elevation of 1816 metres above sea level, the sweeping panorama of grassy slopes and dense *shola* forests in the folds of the hills is overwhelming. The origin of the grasslands in the montane regions of southern India has been the subject of considerable debate. Some botanists believe that the prevalence of frost during winter limits the growth of woody plants and thus the grasslands are really the best adapted natural vegetation. Others have conjectured that early human settlers cleared the *shola* forests and regularly set fire to the vegetation, thus creating and maintaining the grasslands. The studies conducted so far, including some of my own, have shown that natural grasslands did exist over twenty thousand years ago.

The valley between the central hills was once covered with dense rain forest. Sanderson (1878) describes it thus in his book *Thirteen Years Among the Wild Beasts of India*:

> This deep, forest-encumbered valley is a tract of great interest; and there are many places which I have penetrated where, I believe, other European foot never trod. Wild swamps there are where the strangest forms of vegetation are seen, some found nowhere else in the hills. . . . Aged trees of huge dimensions, whose ponderous arms are clad with grey moss and ferns far out to their points; tough, gnarled, leafless creepers, as thick as a child's body . . . the whole is a damp, hoary forest, sacred as it were to the first mysteries of nature.

Very soon European foot trod rather heavily on this seemingly impenetrable jungle. Beginning in 1888 these forests were cleared by British coffee planters. The oldest of these estates, Attikan Estate, was opened by Randolph Hayton Morris. In 1895 R. H. Morris was gored by an injured gaur bull while he was out hunting, but he survived despite losing one lung in the encounter. The Morris family continued clearing the area for coffee, eventually covering about a thousand hectares of the hill slopes and the valleys. R. H. Morris's second son, Ralph Camroux Morris, was a well-known sportsman-cum-naturalist

Once common in southern India, tuskers such as this are now a rare sight after the wave of ivory poaching.

A herd feeds on the profuse fresh green grass that has sprouted after the recent rains.

Feeding avidly on mud at a salt lick.

who recorded many interesting observations of the flora and fauna of the Biligirirangans in the *Journal of the Bombay Natural History Society*. R. C. Morris left the country in 1955 after selling the estates to Indian planters.

Only remnants of the rain forest survive. In one of these patches a huge *Michelia champaca* tree, locally known as the *Dodda Sampige*, is considered sacred by the Sholaga people. They believe that stones appear spontaneously around plants, which then have to be nurtured and worshipped. Not only large trees but even young saplings of many species are given protection. Such sacred trees or groves abound in many regions of the country, a primeval form of human association with nature, very much relevant in the modern context of depleting biological diversity.

The lower hills to the west are covered with moist deciduous forest, and it is here that elephants are plentiful. Conspicuous among the canopy trees are *Terminalia crenulata* valuable for its timber, *Terminalia bellirica*, the rosewood *Dalbergia latifolia*, the jamun *Syzigium cumini*, the silk cotton *Bombax ceiba, Schleichera oleosa, Grewia tiliaefolia* and various species of fig *Ficus*. The understorey features elephant-broken stems of *Kydia calycina, Grewia* and *Helicteres isora*, and 10-foot high grass *Themeda cymbaria*. The vegetation turns to dry deciduous forest as one goes westwards or southwards, with *Anogeissus latifolia* increasingly dominating.

Until recent times the *podus* or tiny hamlets of the Sholaga people dotted these forests. The Sholagas are the earliest inhabitants of the region; they may well have moved in over two thousand years ago. They were mainly food gatherers living on honey, fruits, tubers, the shoots of forest plants, and the meat of animals such as the hare and deer they trapped or found dead. In the past they were engaged in *kumiri* or shifting cultivation. These sites have now reverted to jungle. Beginning in the 1960s most of the scattered Sholagas were brought together by the government into permanent settlements, the largest of which is at BRT Betta where the ancient temple is situated.

These tracts were highly malarial until a few decades back. This is reflected in the high incidence of sickle-cell anaemia among the Sholagas. A person afflicted with this genetic defect has abnormally shaped red blood cells, especially when oxygen is in short supply. One who is homozygous (has two identical genes) for the sickle cell gene usually dies at a young age. But one who is heterozygous (that is, having one sickle cell gene and one normal gene) has a better chance of resisting malaria and surviving compared to a person who is normal homozygous (having two perfect genes for red blood cells). Apparently the malarial parasite does not multiply well in the red blood cells of those heterozygous for the sickle-cell trait. In this case a 'defective' gene actually confers an advantage to an individual who possesses it in a single dose.

The forests here and over most of the Western Ghats, because of their unhealthiness, have never been very attractive for permanent settlement to people from the plains. The tribes living here have adapted to the conditions over the centuries. H. Sudarshan, a medical doctor working in the Biligirirangans, estimates that over a fourth of the Sholagas carry the sickle-cell gene in a single dose, while another two per cent carry a double dose of the gene. During the 1950s the government launched a pesticide war on the malarial mosquito. Once relatively freed from this disease, the stage was set for people from outside to populate the interior hill forests.

Even a century ago the plains had been almost entirely put under the plough. To the west the Biligirirangans drop steeply to the Mysore plateau, a dry flat land of 750 metres elevation. At the foothills only a narrow strip of dry jungle remains. This was the case, though the jungle then extended somewhat further, even in 1874 when Sanderson successfully organized for the first time in the south the capture of elephants by the *kheddah* method of driving them into a stockade. He achieved this along the river Honhollay, now known as the Nirdurgi.

About a century prior to this, Hyder Ali, the ruler of Mysore, had made an attempt to capture an entire herd in the Kakankote jungle but had failed. He prophesied that no one would succeed in the future and apparently pronounced a curse upon anyone attempting to do so, even recording this on a stone still standing in the jungle. Hyder's curse seemed to work initially because Sanderson failed in his first attempt to capture a herd in November 1873. But he persisted and succeeded in his next attempt, made the very next year.

> On May 5th a large herd of elephants came down the hills into the low-country jungles. . . . On the 9th of June . . . we occasionally heard the trumpet of the elephants, fully three miles

> distant, as they fed and disported themselves about the river. . . . At 9 a.m. (on 10th June) we started. . . . It was past mid-day before we got all the elephants into the cover, and not a moment's rest did any of us get till 11 p.m. . . . The river was an advantage, as the elephants had easy access to water. The lurid glare of the fires, the gaunt figures of the lightly-clad watchers, their wild gesticulations on the bank with waving torches, the background of dense jungle resonant with the trumpeting of the giants of the forest—formed a scene which words are feeble to depict.
>
> By 11 p.m. the defenses were thoroughly secured . . . that the elephants could not now escape was certain, unless indeed they carried some of our barricades. . . . The men differed as to their number. I had seen about twenty; some declared there were fifty, but I could not believe this. . . .
>
> On the morning of the 17th, everything being in readiness for the drive . . . we succeeded in moving them through the thick parts of the cover with rockets, and soon got them near to its entrance. . . . The elephants, however, when near the entrance, made a stand, and refused to proceed, and finally, headed by a determined female, turned upon the beaters. . . . I intercepted them, and most of them hesitated; but the leading female, the mother of the albino calf . . . rushed down upon me, as I happened to be directly in her path, with shrill screams, followed by four or five others . . . when within five yards I floored her with my 8-bore Greener and 10 drams . . . the shot was not fatal, as the head was carried in a peculiar position, and the bullet passed under the brain. The elephant fell to the shot, almost upon me . . . and I gave her my second barrel, which must have penetrated to the region of her heart, as a heavy jet of blood spouted forth when she rose. . . . For some moments she swayed from side to side, and then fell over with a deep groan, to rise no more. . . . After a short pause . . . the last of the herd passed in with a rush. . . . C who was perched on a high branch of the gate-tree cut the rope, and amidst the cheers of all, the valuable prize of fifty-three elephants was secured to the Mysore government.

A similar scene was enacted during the last *kheddah* at Kakankote in January 1971. The leading matriarch who refused to enter the stockade was shot dead before the remaining forty-one elephants were driven in. By this time the conservation movement in the country was fledging. The gun was giving way to the camera. An uproar over the cruelty involved in capturing elephants finally led to the *kheddah* being given up. In any case the Kakankote site was soon submerged due to the construction of a dam across the Kabini river. Between 1874 and 1971 just under two thousand elephants were captured by *kheddahs* in the Mysore forests, chiefly in Kakankote and in the Biligirirangans.

Not far south of the site of the *kheddah* at the foothills of the Biligirirangans is the village of Punjur. During Sanderson's time there was only one family, that of 'old Bommay Gouda', in the village; today the population is about a thousand. Punjur lies at a strategic position for both elephant and man. From Punjur southwards a ten kilometre long valley leads to Hasanur.

The Araikadavu stream flowing northwards along this valley is a favourite haunt of elephants during the dry months. Stately white-barked *Terminalia arjuna* trees line the banks along with mango, jamun and bamboos. Elsewhere in the valley the vegetation is dry deciduous forest with a dense undergrowth criss-crossed by elephant paths with open glades in places. Umbrella-shaped *Acacia leucophloea, Acacia sundra, Anogeissus latifolia*, flame of the forest *Butea monosperma, Ziziphus xylopyrus* and wood-apple *Feronia elephantum* are the common trees, while the undergrowth is a thorny tangle of *Acacia pennata, Capparis sepiaria* and the exotic weeds *Lantana camara* and *Chromolaena odorata* among other shrubs. Midway across the valley at Karapallam, the border between Karnataka and Tamil Nadu states, a patch of gregariously growing *Acacia suma* is a target of considerable damage by elephants.

It was near Karapallam that Sir Victor Brooke and Colonel Douglas Hamilton in July 1863 shot a bull elephant with one of the longest tusks ever recorded. Brooke's account has been recorded by Sanderson.

> As we arrived on the ridge overlooking the valley where the elephants were, we heard the crackling of bamboos, and occasionally caught the sight of the back of an elephant. . . . With marvellous silence and quickness the huge beasts marshalled themselves together, and by the time they appeared on the more open ground in the centre of the valley, a mighty cavalcade was formed which, once seen, can never be forgotten.

> There were about eighty elephants in the herd. Towards the head of the procession was a noble bull, with a pair of tusks such as are rarely seen in India. Following him in direct line came a medley of elephants of lower degree—bulls, cows, and calves of every size, some of the latter frolicking with comic glee, and bundling in amongst the legs of their elders with utmost confidence. At length the great stream was, we believed, over, and we were commencing to arrange our mode of attack, when that hove in sight which called forth an ejaculation of astonishment from each one of us. Striding thoughtfully along in the rear of the herd, many of the members of which were, doubtless, his children, and his children's children, came a mighty bull, the like of which neither my companion, after many years of jungle experience, nor the two natives who were with us, had ever seen before. But it was not merely the stature of the noble beast which astonished us, for that, though great, could not be considered unrivalled. It was the sight of his enormous tusk, which projected like a long gleam of light into the grass through which he was slowly wending his way, that held us riveted to the spot.

Brooke then describes how he shot the bull and bagged its tusks. Its left tusk was diseased and measured only 3 feet 3 inches long, but its right tusk measured an enormous 8 feet and weighed 90 pounds. At that time it was probably the longest and heaviest known Indian elephant tusk. Since then only two or three elephants have yielded bigger and heavier tusks in India.

The Araikadavu valley from Punjur to Hasanur and beyond to Dimbam has witnessed the passage of people for many centuries. This and the Talamalai plateau leading down to Gazalhatti further west were natural passes between the Mysore plateau to the north and the Coimbatore plains to the south.

The early inhabitants of the hill forests would have been hunter-gatherers and shifting cultivators, organized into endogamous groups enjoying a certain degree of autonomy. They certainly would have interacted with more advanced peasant societies which in turn owed allegiance to the various dynasties which reigned to the north and the south. Since the beginning of the Christian era this region has come under the influence of the Pandya, Chola and Chera dynasties which ruled the south, the Gangas who ruled over most of Mysore, the Rattas, the

Kadambas, the Vijayanagar kings and the Muslim rulers whose seats of power were in the north. The Cheras who ruled the region known as Kongu land, which included the Coimbatore plains and surrounding hills, for the better part of the first ten centuries had their capital for sometime at Skandapura near the Gazalhatti pass, at the confluence of the Bhavani and the Moyar rivers.

During the late nineteenth century the Muslim rulers of Mysore, Hyder Ali and his son Tippu Sultan, held sway over this region. They frequently marched their troops through this area during their incursions against the British and on southwards into the Coimbatore plains and the Carnatic. Tippu even posted guards to look after the forest wealth, especially the fine sandalwood found here. Their legacy can still be seen in the roads lined with banyan trees, in the route of the bridle path from Talamalai down to the Gazalhatti pass, in the crumbled bridge across the Moyar river at the foot of the pass and, further west, in the ruins of a fort overlooking the Moyar gorge in the Mudumalai forest.

The Mysore generals and the British used Lambadis, a gypsy tribe which migrated from the north of the country as carriers of supplies for their troops. Today, the Lambadis settled in the village of Kolipalya to the west of Punjur are agriculturalists. Lambadi women are conspicuous in their colourful dresses and metal ornaments, although the ivory bangles they once sported are hardly seen these days.

After Tippu was overcome by the British in 1799 the throne was restored to the Wodeyar dynasty, but Mysore for all practical purposes was governed by the British. The Wodeyar princes of Mysore maintained their private hunting preserves, among them the forests of Bandipur and the Biligirirangans. The contiguous forests of Satyamangalam to the south came under the British East India Company. In 1856 the British organized the forest department of the Madras Presidency. The Satyamangalam and Talamalai forests were among the earliest declared reserves.

To the east of the Hasanur–Dimbam tract the dry hills are dominated by stunted trees, chiefly *Anogeissus latifolia*, and the shrubby palm *Phoenix humilis*. The Badagas, who migrated here after the fall of the Vijayanagar kingdom during the sixteenth century, cultivate the plateau settlements of Kotadai, Devarnattam and Mavallam. They also maintain large stocks of cattle in *pattis* or pens inside the forest. Further north the village of Gaddesal is a Sholaga settlement nestled picturesquely at the foothills of the central hill range.

To the west of the Araikadavu valley the hills are again covered with dry deciduous forest of *Anogeissus latifolia* with *Phoenix humilis* and grasses such as *Themeda triandra* and the lemon scented *Cymbopogon flexuosus*. Small patches of semi-evergreen forest are conspicuous in places where the soil favours their maintenance. Gregarious stands of the aromatic, resin-yielding *Shorea roxburghii* are also conspicuous on the hill slopes. The hills drop on the west to the Talamalai plateau where four large enclaves—the villages of Chikkahalli, Neydalapuram, Mudianur and Talamalai—are separated by narrow strips of jungle. These villages, as well as Punjur and Hasanur, are settled mainly by Lingayats, the most populous of all the peoples in the study area. The Lingayats are followers of Basava, a twelfth century reformer who repudiated orthodox brahminism. They are mainly agriculturalists and cattle breeders.

South of Talamalai the hills drop steeply to the Moyar river valley or the Gazalhatti pass. Here east meets west; that is, the chain of hills collectively known as Eastern Ghats meets the Western Ghats at the Nilgiri hills. The valley at an altitude of only 250 to 300 metres is hot; it is also very dry, save for the perennial Moyar, being sheltered by the hills from the rain-bearing clouds. Silvery grey *Gyrocarpus jacquini*, umbrella-shaped *Acacia leucophloea*, heavily browsed *Albizia amara*, bilobe-leaved *Hardwickia binata*, cactus-like *Euphorbia* and a host of thorny plants predominate the semi-arid landscape.

At the confluence of the Moyar with the Bhavani a dam impounds the waters for irrigating the Coimbatore plains. An evening drive along the foreshore of the reservoir is invariably rewarded with the sight of dozens of blackbuck and, during the right season, herds of elephants coming in for a drink. With a little more luck one might even see the striped hyaena, a characteristic animal of these dry jungles. During the late nineteenth century a cheetah had been shot at this very place. In 1948 the last wild Indian cheetah was shot and the species became extinct in the country. Another animal to have disappeared from here which was once not uncommon in these jungles is the wolf.

Twenty kilometres upstream, past the villages of Hallimoyar

and Tengumarada, the Moyar emerges out of a three hundred metre deep gorge at Mangalpatti, which was the southwestern limit of my main study area. Further west, elephant trails lead up to the Sigur plateau, which again has only dry jungle as it lies in the rain shadow of the Nilgiri massif. The plateau is dotted with agricultural settlements and cattle and buffalo *pattis*. The human communities include Irulas, Badagas and Kurumbas. Human occupation of the plateau is ancient, as can be deduced from the cairns, barrows and kistvaens, the burial sites of a megalithic culture found here, many of which were constructed between AD 100 and 1100. There are also numerous dolmens and *virakals*, dating after AD 1200, sculptured with figures of people and animals, depicting scenes of battle and hunting.

The village of Masinagudi, at the western end of the Sigur plateau, lies at the transition from the dry thorn forests to the extensive deciduous teak forests of Mudumalai Sanctuary. Formerly known as Masinahalli, this village was apparently once the capital of a chieftain ruling the Wyanad region during the sixteenth century. This was a much cultivated and populous tract in earlier times, but wars and famines during the eighteenth and nineteenth centuries contributed to its decline. During the twentieth century Masinagudi again acquired more importance with the coming of hydro-electric projects in its vicinity. The increasing human and livestock populations have put considerable pressure on the forests in recent years.

Unlike the Biligirirangans, the deciduous forests of the Nilgiri foothills have plenty of broad-leaved teak *Tectona grandis* trees. The rainfall increases as one goes westwards and southwards; consequently the vegetation changes to moist deciduous forest. In the dry forest *Anogeissus latifolia* dominates along with *Tectona grandis*, but in the moister parts these give way to *Terminalia crenulata, Kydia calycina* and *Lagerstroemia microcarpa*. *Vyals* or swamps bordered by huge bamboos *Bambusa arundinacea* extend like tentacles through the moist forests; hence the name Wyanad (actually *vyalnad* or the land of swamps) for this region. Chettis cultivate many of these swamps today.

The Kurumbas are among the earliest inhabitants of these forests. They are very similar in appearance and habits to the Sholagas. Like the latter, they were also primarily food gatherers and hunters of small animals and in addition carried on shifting cultivation. Currently, like most other tribal peoples, they are at the crossroads with more technologically advanced cultures.

The deciduous forests extend northwards through Bandipur to the banks of the Kabini river, the site of twenty-four *kheddahs*. The dam across the Kabini has reduced the forest corridor here to only five to six kilometres across. Gliding along the backwaters of the Kabini in a coracle, a circular boat made of bamboo and buffalo hide, one can count over a hundred elephants along the banks on an April afternoon. North of the Kabini, the deciduous forests continue into the Nagarhole National Park, probably the best place today in southern India for seeing a tiger or a leopard.

The complex of Nagarhole, Bandipur, Wyanad and Mudumalai reserves, covering over 2000 square kilometres, is one of the finest places anywhere in Asia for seeing, studying and photographing wild elephants. Taken along with the adjoining Sigur plateau and the ranges of the Eastern Ghats, including the Biligirirangans, Bargur, Madeshwaramalai, Dharmapuri, Kanakapura and Bannerghatta extending to the suburbs of Bangalore city, a total forest area of about 10,000 square kilometres, this harbours one of the largest known Asian elephant populations of 5000 to 6000 individuals. As late as the 1960s the area was even more extensive, as Mudumalai was then linked to the wet evergreen forests of Nilambur and the western slopes of the Nilgiris. However, the planting of tea to the west of Gudalur has broken this link, although some elephants still manage to move across the plantations. This fantastic diversity of habitat types, from semi-arid scrub receiving just 40 cm. of rainfall annually to monsoon evergreen forests enjoying 400 cm. precipitation, I believe holds, or should I say held, the key to the elephant population.

As I explored my study area I also established contact with farmers in the villages. The village centre, usually a large *arasu* or *peepal (Ficus religiosa)* tree where the menfolk relaxed in the shade after their morning toil, was a good place to make contact. The average Indian peasant is a simple, friendly and hospitable soul. At the same time he was also very curious about what this 'officer' was doing in his God-forsaken village.

'*Saar*, you are a native of?', he would enquire.

'India', I would reply.

'*Saar*, what caste are you from?', he would try again.

'Indian'.

'What is your salary, *saar*?'

'Enough to keep me alive!'

(The importance or otherwise of an official largely depended on his 'basic salary'. I heard the story of a Collector, the head of a district, who had to admonish an official from another department for not attending meetings presided over by him. The official used the excuse that his basic salary was higher than that of the Collector!)

'*Saar,* do you have the rank of a DFO (District Forest Officer) or a Collector?'

If I came in a jeep, I had to be a government official. In a hierarchy-conscious society, it was a little difficult explaining to him that I was not in the government, but that I was interested in finding out why elephants raid crop fields and how much damage they caused.

Hope springs eternal. Decades of governmental apathy to his woes had not completely extinguished the Indian farmer's spirits. Perhaps I would plead to the government on his behalf. Perhaps he would be compensated for his losses. I assured him that I would certainly present the results of my study to the government and recommend that suitable measures be taken to keep elephants away from his doorstep.

In March and April most of the fields are fallow. Only at Talamalai were some farmers still cultivating *ragi* (or finger millet) using water from wells. I received complaints that Vinay, the bull with the broken right tusk, had visited these fields and pushed over coconut trees. About a month earlier, on 10 February, Vinay had claimed the life of Rama who had been guarding the millet fields. I also saw some damage to banana and mango orchards at Neydalapuram and to coconut trees and sugar-cane fields at Chikkahalli. With the exception of a family herd of four elephants, the culprits were lone bulls.

The villagers were, by and large, enthusiastic about informing me of raids. They, of course, had a tendency to greatly exaggerate the extent of the damage suffered but, provided I inspected the fields personally, I found that I got very useful information on the frequency of raids in the villages, the quantity of crops consumed, the loss to the farmers, and whether bulls or herds were responsible.

Since the seasonal movements of elephants could be expected to be related to their feeding habits, I also had to get an idea of when and how far the elephants in the study area moved. The seasonal densities in the different habitat zones I had defined would provide clues to their movement patterns, but I still had to relate this to individual elephant bulls and herds. I did not have the facilities to put radio-collars on elephants or to mark them in any way. So I set about doing the only other thing possible—to identify as many elephants as I could and hope that I would be able to see at least some of them regularly.

The males were fairly easy to recognize; most of them were tuskers and their tusks were of various shapes and sizes. The first bull I registered in the study area was near Dimbam. It had thick but short tusks and I called him Tippu after the dapper Mysore general, who undoubtedly would have encountered elephants at the same spot two centuries ago.

Then there were the Karapallam bull with a third of its left tusk broken, Dodda Sampige, the oldest bull I encountered, having thick long tusks with a foot of the right one broken off, young Biligiri, to whom I referred earlier and, of course, Vinay, with his right tusk broken near the lip. One tuskless bull I simply called Makhna haunted the Araikadavu valley. By the time I finished my study I had identified twenty-four adult bulls in the area. There were others which I could not register for want of a good photograph.

It was a far more difficult task to recognize the female elephants. For one thing they do not have tusks, although a few have rudimentary tushes which are visible externally. Their ears do occasionally have cuts and holes, but these are not as common as in the large-eared African elephant. Cows above thirty years of age gradually develop a fold on the top of their ears (in bulls this folding may begin by the time they are twenty). By a combination of ear characteristics, the appearance of tushes and the presence of warts on the body, I managed to identify a number of cows. The composition of the family group was another aid, but this could not be always relied upon because the groups did sometimes split into smaller units. Eventually I registered twenty-four different family groups comprising 212 individuals.

One of the aims of most ecological studies of a species is to find out the trends in the population number, whether it is increasing, decreasing or stable. This was true in my case. I wanted to know how the elephants were faring in their interactions with people. Crop raiders were sometimes killed and significant numbers of males were being poached for tusks. To study the dynamics of the elephant population, I first had to estimate the ages of the elephants, particularly those in family groups, by simple field methods. I also had to establish when cows began reproducing, how often they gave birth and when they stopped reproducing. In addition, I had to know how many elephants died each year and their ages at death.

Ian Douglas-Hamilton, who studied elephants at Lake Manyara National Park in Tanzania, fabricated a stereo photographic device to measure the shoulder heights of elephants, and related these to the growth curves for height against age developed by Richard Laws. I too had to age elephants by measuring shoulder heights but wanted to do so by a simpler method. Although I did make a similar device, I found it rather impractical to regularly carry around.

One morning in March, just two weeks after I began work, I got the answer. Driving from Hasanur to Dimbam I saw an elephant herd coming up a valley. They slowly came along the dry nullah, pausing occasionally to reach out for a clump of grass or an overhead branch, or to rest in the shade of a mango tree. The largest elephant, a cow, had sunken cheeks and ears folded at the top. Realizing that they might cross the road after a while, I positioned myself some distance away from their path which ran amidst the dense *Lantana* bushes. As the elephants crossed one by one I photographed them; then it suddenly stuck me that if I measured the distance from my position to where they had crossed, I could measure their image sizes on the photo negatives and calculate their heights by incorporating these figures and the focal length of the lens I was using into standard mathematical equations. Fortunately, I was carrying a measuring tape for estimating the damage to crop fields and, as soon as it was safe enough, Setty and I measured the distance, which was 47.8 metres. All that I had to do was to develop the film, make magnified prints to determine the object size more accurately, and then calculate the heights of the elephants. To aid in the calculations I calibrated a pole of known height using the same lens and made corrections for lens distortion.

I still had to establish the relationship of shoulder height to the age of the elephants. Although some growth curves existed for Sri Lankan elephants, I was not sure if these could be applied to the elephants in my study area. For this I needed to measure captive elephants that came from southern Indian populations.

The first family group I photographed near Dimbam and classified in this manner was, by coincidence, also the one I came to know the best. It was a joint family of nine individuals, led by an old matriarch I estimated to be over fifty years old, and encompassing three generations. Standing 256 cm. at her shoulder she was also the tallest cow I ever registered. She had two mature daughters twenty to thirty years old, one immature daughter about fourteen years of age and two sons aged nine and three. The mature daughters had children of their own; the older daughter had a young twelve years old cow and a son of three to four years of age, while the younger daughter had a calf less than a year old.

During several encounters with this family I noticed that the matriarch was always calm and composed, leading her family with a quiet yet majestic confidence. I named her Meenakshi, after my paternal grandmother, a woman I remember as having a strong, commanding personality.

There were many other herds, which could be usually seen either along with or close to Meenakshi's family. There was another old cow along with her children and grandchildren. She was often seen along with Meenakshi's family and so I presumed the two were sisters, Meenakshi being the older one. This cow was distinctly shorter than Meenakshi and she reminded me of Tara, the oldest captive elephant in the camp at Mudumalai.

Another family of four sisters, all of them tall and lanky, leading their eight children ranging in age from thirteen years to less than a year, was called the Mriga family. In the traditional system of classification, a *koomeriah* (or thoroughbred) is a stocky, barrel-shaped and well proportioned animal, one that belonged to the highest caste, while a *mriga* (or deer in Sanskrit) is a more delicate, lanky animal. There was High Head, so named because of the distinctive manner in which she carried herself, and her group of seven elephants. And there was Kali, a cow with a large swelling on her forehead that made her fierce-looking (Kali is a particularly angry Hindu goddess). She was part of a family which included her three mature sisters and their seven children when I recorded them all in September 1981.

Other elephant herds ranged to the north in the Biligiriran-gans. I never saw any of their identified members either near Dimbam or in the Araikadavu valley between Punjur and

Hasanur. One of the matriarchs was unmistakable due to the large U-shaped tear in her left ear. Her immediate family included an eleven year old daughter, a seven year old son and a calf of under one year, but she also moved in the company of other elephants which included at least three adult sisters and their children—a family of twelve elephants in total. Although the matriarch's temples and cheeks were sunken, indicating that she was in poor condition, she was a relatively tall elephant. I named her Champaca after the specific name of the *Dodda Sampige (Michelia champaca)* tree in whose vicinity she ranged.

Champaca's family and many others in the northern Biligirirangans constituted Clan 1. Meenakshi's clan occupying the Araikadavu valley further south was Clan 2. Apart from these two clans there were at least three other elephant clans in the study area. These moved out of the area for a good part of the year and I positively identified few of their members.

Within three months of beginning work I registered ten families, which constituted 40 per cent of my final tally. Considering the difficulties in observing elephants under forest conditions this seemed a good start. During this time there was relatively little crop raiding and, hence, I could afford to spend most of my time in locating elephants.

CHAPTER 3

The Renewal of Life

Oh! Lord of the hilly tract, where the elephant which ate the bamboo along with its kin vanquished the tiger . . . and after wiping the blood from its tusks, walked slowly and magnanimously, full of pride, and got united with its mate and lay asleep with the bees hovering around its rut.

Tamil Sangam Literature (1st–5th century AD)

It was late in the morning of 15 March 1981, the sun very nearly directly overhead. Along the Hasanur–Dimbam road the hill slopes were tanned by the blazing sun, their surfaces scorched in patches by the fires that had already swept through them. Down in the valley a narrow belt of green trees clung tenuously to the banks of a dry nullah, surely the most refreshing of sights for a traveller or an elephant-watcher on a hot and dusty March day. One of the trees, a large mango, was especially appealing. As I looked down longingly at this green oasis, already sensing the cool breeze its leaves had to offer, I noticed that someone else had similar ideas. That someone was Tippu and he was in full musth. The only thing I could do was to remain where I was in the merciless heat, to take out my notebook and begin writing.

11.15 a.m.: Tippu is standing under the tree with his trunk resting on the embankment. He is quite still, almost statue-like.

11.23 a.m.: He places his short, thick tusks on the embankment and presses down gently. His right temporal gland exudes a fluid which flows down his already darkly stained cheek. Almost immediately he lunges forward at a nearby bush, thrashing it for some twenty seconds. After that he is quiet again.

11.30 a.m.: He continues to stand still, but now I notice that his penis is partially erect and flaccid intermittently.

11.44 a.m. : He now places only his trunk horizontally on the embankment and presses gently for a few seconds. Stands still for sometime. Places his trunk sideways and presses. Alternates between pressing and withdrawing his trunk for the next ten minutes.

11.55 a.m.: Tippu sits down on the stream bed in sternal recumbency (that is, on his belly).

12.15 p.m.: Turns over on to his side. He is asleep by now. I find that my camera has run out of film. I leave for Hasanur to pick up a roll.

1.24 p.m.: When I return to the spot I find that Tippu has got up, probably just a few minutes earlier, and is moving behind the *Lantana* bushes. Slowly he moves in a northerly direction through the valley, and up the opposite hill along a gently sloping path, at a speed that for an elephant can only be described as a snail's pace.

1.50 p.m.: I can still see him on the hill slope.

When I began my work in late February I frequently saw elephants in the hilly tract between Hasanur and Dimbam. Tippu was the first adult bull I recorded. He was rather short and stockily built with a pronounced bump on his forehead. His tusks were also very short but quite thick; I put his age at about forty years.

The first time that Setty and I saw him, he was half hidden in the dense undergrowth, feeding patiently on the bark of acacia shrubs. After half an hour he suddenly noticed us and came up to the road with his head held high in defiance. Setty ran around to the front of the jeep, climbed up on the bumper and crouched low in an attempt to hide from Tippu. I tensed for a moment in the driving seat, looking back at the elephant, not daring to start the jeep and risk running over my tracker! After Tippu had crossed the road without any further threats, I began my first lesson for Setty by explaining to him that he should get into the vehicle rather than sit in front of it when confronted by an angry elephant.

Tribal scouts are indispensable for moving about in elephant country; the best among them have the eyes of a hawk, the nose of a bloodhound, the ears of a fox, and the analytical powers of a Sherlock Holmes. But it is also important for researcher and tracker to work in co-ordination; to have a perfect mutual understanding of how to move when faced with a crisis. A foolish move by one can endanger the life of the other. There were many details of field biology that one could learn from them—how to identify a tree from the colour of its bark or how to tell in which direction an elephant had moved by looking at its footprint. But there were also many things one had to teach them. Setty, for instance, surprisingly did not know that some male elephants do not possess tusks. He mistook the first *makhna* we encountered for a female elephant. Anyhow, Setty and I built up a good rapport quite early as we went about our business of locating and observing elephants.

Tippu seemed to be the only big bull ranging over the hills to the south of Hasanur during this time of the year. We did, however, encounter many herds there. Meenakshi's family was certainly among them, along with many other families that were probably related to her. Soon, by the end of March, the focus of our work shifted north of Hasanur.

My little bungalow atop a hillock afforded a marvellous view of the surrounding hills and jungles. After the sun sank behind the hills to the west I would sit quietly outside listening to the calls of the elephant herds. Every evening a jackal would emerge from the jungle behind my bungalow and cautiously make its way past the front door to the road below in search of village fowl. A black-naped hare would quickly hop over from the bushes in front of the bungalow to a stone in an open patch, crouching briefly to make sure that no predator lurked nearby, before making a dash to the jungle behind. The stars lit up the darkening sky, sparking an infinite variety of patterns in the human imagination, with the Milky Way stretched out as a wispy cloud across the star-studded dome. The trumpeting of the elephants reverberated across this amphitheatre, providing the final climax to a grand orchestra that no city-dweller is usually privileged to experience.

During the time in which the elephants settled into the Araikadavu valley the nights were filled with deep rumblings and the crackling of dry brush as the huge beasts moved to and fro through the jungle behind my room in search of the most suitable locality. This provided me with a good opportunity to observe elephants within a short distance of my bungalow.

Two kilometres beyond Karapallam on the road to Punjur, a small pond close to the road was a favourite haunt of elephants during the dry season. It was also a very convenient location for me to observe and photograph them for my identification file. I would park my jeep under one of the tamarind trees and either climb up it or wait behind a bush near the pond for the elephants. Setty would scout the area to see if any elephants were close by.

Early April was a very rewarding period at the Karapallam

pond. I was able to classify Meenakshi's family and many others related to her. On the morning of 9 April I had parked myself in the maidan near the pond. At 10.00 a.m. the splash of water heralded the arrival of elephants. Moving behind a bush I saw a large herd already in the pond.

> The elephants are in two sub-groups, one of ten and another of five animals. They have come from the jungle to the west of the pond. Standing knee-deep in the water, they reach out with their trunks sucking in the liquid and blowing it into their mouths. Soon they spread out, sinking into the water, the youngsters imitating the elders or clambering over their backs. After ten minutes they quickly turn back and climb out of the pond.
>
> One of the adult cows with folded ears has a wart on her right flank. She does not have any hair at the tip of her tail (this was Meenakshi). She stands beneath a bamboo clump and a young tusker about eight years old comes up and rubs his bottom against her left side. Another old cow with folded ears, but distinctly shorter, is also with the group of ten elephants (she was the elephant I later named Tara).

By 10.15 a.m. all the elephants had disappeared into the jungle. I moved over to the bridge on the road, waiting for them to cross. Tara appeared first with two of her youngsters followed by Meenakshi and six others. As they crossed the road I photographed them at a distance of 51.5 metres. The other five elephants crossed over a few minutes later. I went back to the pond.

> 11.00 a.m.: Another herd of nine elephants has entered the pond. The Karapallam Bull has also entered along with them.
>
> 11.05 a.m.: The herd gets out of the water. Some of the elephants rub themselves against a slanting tree. The Karapallam Bull remains in the pond.
>
> 11.15 a.m.: The herd crosses the road. I photograph it at a distance of 53 metres.
>
> 11.30 a.m.: The Karapallam Bull suddenly gets out of the pond. Three more elephants have come from the north. The Karapallam Bull goes up to a sub-adult cow and jabs her with his left tusk. He then leaves her alone. The cow visits the water briefly before rejoining the other two elephants.
>
> 11.40 a.m.: The herd of three crosses the road.

The procession of herds to the Karapallam pond continued for the next three hours, after which I returned to Hasanur.

Over the previous few days I had been observing the Karapallam Bull at the pond on various occasions. He was not very old, perhaps twenty to twenty-five years of age. He was in musth at this time. On 5 April I recorded:

> 10.20 a.m.: The Karapallam pond has very little water. A lone bull is standing at the edge of the water, sucking the slushy liquid with his trunk and spraying it on to his sides. His right tusk is medium-sized and sharp-edged. A third of his left tusk is broken. His temporal glands are active. Fortunately, the wind is in my favour. There is a strong smell, coming from the musth fluid?
>
> 10.30 a.m.: He is slowly moving to a nearby tree. Rubs himself against the tree trunk. He even rubs individual parts of his body such as his hind legs and tail.
>
> 10.40 a.m.: I lost sight of him as he has moved into the bushes.
>
> 10.50 a.m.: A tusker suddenly burst out of the bushes, followed by the Karapallam Bull in hot pursuit. Setty and I turned to run but fortunately they veered away in the opposite direction out of sight.

A few moments later a herd of seven elephants emerged out of the bushes and crossed the road. There was no doubt that the musth Karapallam Bull was asserting his dominance over the rival who had managed to sneak up to the herd while he was away. I could not record sufficient detail to identify the other bull because the entire sequence was over in a flash.

Even if I did have sufficient time I would not have been able to get a good picture of the bull. My telephoto lens had broken, the optical unit having come loose from its helical focusing mount after a particularly rough ride down to Punjur from the Biligirirangans. The past few days had been very frustrating indeed. The Karapallam pond was swarming with elephants and all kinds of exciting things were happening, and here I was without the proper equipment to record these events. With my standard lens the objects of interest were just too small. That day I decided to do something about my lens. There was no repairer nearby who could have fixed it properly for me. So I did the next best thing possible. I inserted the optical tube into its mount and looking through my camera viewfinder found that I could see something. There was no way to set the lens aperture but I could make a guess at it by looking through the lens. With this crude contraption, that just stopped short of being tied together with the proverbial shoe-lace, I went ahead and shot several rolls of film. These still remain some of my best pictures of elephants!

The next morning I parked my jeep close to the pond and sent Setty to check for fresh signs of elephants. A soft purr, not unlike that that of a motorcycle starting in the distance, reached my ears. I thought to myself that someone had just finished a late breakfast of hot *dosais* and *idlis* at one of the thatch-roofed restaurants at Karapallam and was moving off.

The second time I heard the same faint sound I casually turned back, to see to my horror the Karapallam Bull walking equally casually towards the pond from the road. The problem was that I was directly in his path. He was very close and there was little time to get away. I frantically signalled to Setty who had not noticed him at all. At first Setty could not understand what my concern was but when he finally saw the bull he sprinted back to the jeep. We moved off in a hurry, expecting a thundering rogue to pursue us, but all that the Karapallam Bull did was to continue walking nonchalantly towards the pond as though we did not exist. It was embarrassing, almost insulting! (Some years later, upon reading the work of Katharine Payne on infrasonic communication between elephants, I realized the true significance of the sound I had heard on that day and was to hear on many other occasions.)

After we had recovered from the initial shock, Setty and I summoned the courage to go back to the pond after the Karapallam Bull had moved over to the opposite side. Musth fluid was pouring from his temporal glands and his broken left tusk was packed with clay, no doubt from goring the soil in an effort to ease the musth flow. He had a distinctly malevolent look, though he actually proved to be quite harmless. Before we arrived there he must have already had a drink and a session of goring the slushy soil at the pond. He now totally ignored our presence, even though we were only a short distance away across the small pond.

10.04 a.m.: The bull enters the pond and lies down in the water.

10.08 a.m.: A magnificent trumpet call is heard from the east, the sound echoing from the hills to the west behind the pond. The Karapallam Bull pays no attention to it.

10.11 a.m.: He climbs out of the pond and begins to squirt the muddy water over his back and under his belly.

10.17 a.m.: He lies down again in the slush. Another bull now appears from the west, heading towards the pond. It seems to be the same one that had been chased away the previous day. When it is about 75 metres from the pond it abruptly stops, puts its trunk up to test the air, turns back and bolts as if pursued by a swarm of stinging bees.

10.19 a.m.: The Karapallam Bull gets up and comes out of the pond. He goes over to a tree and begins rubbing his body.

Since the two bulls had definitely not sighted each other, the subordinate bull which was not in musth appears to have retreated merely upon detecting the smell of the Karapallam Bull's musth. The Karapallam Bull, on the other hand, was not even aware of what had happened. It did not really matter. After all, in musth, he was for the present the lord of the Karapallam pond.

Musth, with its sexual connotations, has been known and described in Asian elephants since ancient times. A captive bull in musth turns roguish and is difficult to handle. On occasion it may even kill its mahout or other bystanders. Nilakantha expressed the musth condition aptly, several hundred years ago, in the *Matangalila*: 'Excitement, swiftness, odour, love, passion, complete florescence of the body, wrath, prowess, and fearlessness are declared to the eight excellences of musth.'

During musth there is discharge of a fluid, which has a strong pungent smell, from the pair of temporal glands situated between the eyes and ears. The stains left by the musth fluid on the cheeks may persist for many days after the discharge has ceased. A bull is usually in musth only once a year, for a period of only a few days or up to three to four months. The duration of musth seems to be determined by the bull's age and health; the older the bull or the better its condition, the longer the duration of musth. Mahouts have traditionally taken advantage of this observation to control musth bulls which might become troublesome or aggressive. A mahout may reduce the amount of food given to the elephant when the first signs of musth appears in order to reduce its duration and intensity.

Bulls show the first signs of musth when they are about fifteen to twenty years old. At this age, the discharge from the temporal glands is relatively scanty. Only after twenty-five years of age do they show a more copious discharge. The first time I saw Biligiri in musth was when he was an estimated sixteen or seventeen years old in January 1983. In the previous two years, I had not noticed any temporal gland activity in him.

Work on captive elephants in Sri Lanka by John Eisenberg and his colleagues during the early 1970s firmly established the connection between musth and reproductive activity. However, considerable confusion about the exact significance of musth continued to persist for some more years in scientific literature. This confusion arose from the fact that in African elephants both bulls and cows show active temporal glands. It is even possible to induce secretion by subjecting elephants to some form of stress; for instance, by deliberately chasing or harassing a herd. Obviously this was not a manifestation of rut or increased sexual activity in the male elephant. Temporal gland secretion in African elephants was believed to be merely a form of communication among elephants when the herd faced some external danger. It was thought that 'musth' had entirely different functions in Asian and African elephants—a sexual function in male Asian elephants and a communication function in African elephants.

The observations of Joyce Poole and Cynthia Moss on African elephants during the late 1970s in the Amboseli National Park of Kenya cleared these misconceptions. They showed that there were two types of temporal gland secretion in African elephants. Both male and female elephants secrete a fluid, called temporin, in response to stress in the environment. This is not to be confused with the true 'musth' phenomenon exhibited by the male African elephant. An African bull elephant in musth shows a copious, regular discharge from its temporal glands and a regular discharge of urine from its penis, which turns a greenish colour, a phenomenon they termed as 'green penis syndrome' (or GP syndrome for short). This is similar to the musth that is well known in the male Asian elephant.

When a bull comes into musth its physiology and behaviour alter markedly. The level of the male sex hormone, testosterone, in the blood, rises dramatically. One study by Sri Lankan veterinarian M. R. Jainudeen and his associates found that the level of testosterone in the blood increases some fifty-fold during peak musth compared to the non-musth phase. The musth fluid also contains testosterone in addition to other compounds. A musth bull is more aggressive towards other bulls, has a dominant status in the social hierarchy and is more likely to associate with a female herd. He steps up his search for oestrous cows, perhaps, moving longer distances. All these factors would naturally provide increased opportunity for a musth bull to mate with oestrous cows. It is, however, not necessary for a bull to be in musth in order to mate successfully.

Although elephants are not territorial animals in the strict sense of the term, a musth bull may 'scent mark" an area by rubbing its temporal gland on trees. This might deter other bulls from venturing into the area, although I have no evidence that this actually happens. Whatever is the exact function of musth, it is clear that a bull coming into normal musth would have increased access to cows for mating and, hence, increased reproductive success. The big bulls in good body condition are also the most intense musth machines. But when they are not in musth, a smaller or younger musth bull can gain dominance in the hierarchy.

What happens when two equally big bulls in an area come into musth at the same time? I have never actually seen this happen (partly because of the scarcity of big bulls in the southern Indian populations), but when it does I would presume that the stage is set for a titanic battle which may last for days. The jungle would then resound with the fierce clash of tusks, the crashing of uprooted trees, the deep groans or shrill trumpets of the two giants, a far more earth-shaking event than any sumo wrestling match.

Such fights, though rare, do certainly occur. I have known of three such fights to the finish during 1981–2 in southern India. The loser in all cases had deep wounds on its body. One bull with a single right tusk died near Masinagudi with a one-foot deep wound gored on its right flank. Another large bull, over three metres tall with a pair of tusks weighing twenty six kilograms each, died in the Anamalai hills after it had been gored on its forehead by its rival, the one and a half foot deep fatal thrust piercing its brain. An oversized pair of tusks is certainly no insurance against defeat.

Female Asian elephants also occasionally secrete from their temporal glands, although the precise significance of this is not clear. I have seen cows secreting when they are in an advanced stage of pregnancy or just after calving. The quantity of secretion is relatively meagre. Perhaps the temporal glands of female Asian elephants also have a communication function, similar to the African species, or have some connection with reproductive activity, but this has yet to be unravelled by research.

Elephants can potentially breed throughout the year, although in regions with a pronounced rainy season, breeding can peak at this time. The protein-rich forage available during the rainy season could possibly stimulate the reproductive system to its peak activity. Research on captive elephants in Sri Lanka indicated that female elephants came into oestrus every three to four weeks on average. This observation was based largely on the behaviour of male elephants and the levels of estrogen hormones in the urine of female elephants.

David Hess and his associates from Oregon showed in 1984 that the oestrous cycle of the captive Asian elephant was in fact sixteen weeks in length. They came to this conclusion by monitoring the levels of the reproductive hormones, progesterone and estradiol, in the blood, on a regular basis. Later work has also shown that a potentially fertile ovulation occurs only once every fifteen to eighteen weeks. It would seem that four out of five 'follicular' cycles, each of three to four weeks, end without actual ovulation, while a fertile egg is released only at every fifth cycle.

In the wild, such a regular pattern of oestrus may not occur. During periods of drought the female elephant may simply not come into oestrus at all and they would not, of course, enter the cycle when pregnant or when they are suckling a young calf. The week in April 1981 when I observed the musth Karapallam Bull, many families of Meenakshi's clan were in the vicinity and regularly visited the pond. He mingled with them but, strangely, did not seem to show much sexual interest in any of the cows, except on one occasion. I found out the reason for this only years later. During 1983 very few calves were born in my study area: this meant that, for whatever reason, few cow elephants had come into oestrus during 1981.

When the female is successfully ovulating there is a certain period before the actual event when she sends out chemical cues to attract potential mates. Volatile compounds or 'pheromones' are released in the urine, which is tested by the male elephant. A bull inspects such urine spots with its trunk and places its tip inside the mouth, where a sensory 'vomeronasal' organ located on the roof tests the chemical signal, an action termed as flehmen. If the result is of interest the bull would try and locate the oestrous cow.

A female elephant also advertises her oestrous condition through chemicals released in the genital tract. A bull may perform flehmen by examining her vaginal opening with its trunk. A cow in oestrus may even solicit this examination as Selvam and I observed one afternoon at the Hambetta swamp in Mudumalai.

> 2.07 p.m.: When we arrive at the swamp there are two herds of elephants, one of four and the other of fourteen animals, separated by more than 100 metres of grassland to the west of the lake. They are grazing.
>
> 2.45 p.m.: Another family of five elephants enters the water from the south of the lake. There are two adult cows, two juvenile males and one unsexed juvenile. The entire group bathes in the lake.
>
> 3.00 p.m.: A young adult bull, fifteen to twenty years old, comes into the grassland from the forest to the west. Its temporal glands are mildly active. The bull joins the group of eighteen elephants and begins feeding.
>
> 3.10 p.m.: One of the adult cows in the family of five puts her trunk up in the air. The herd now swims across the lake towards the other elephants. When they near the shoreline they literally rush out of the water. The bull comes up to the water's edge. The adult cow goes up to the bull, who puts his trunk between her hind legs and places it in his mouth.
>
> 3.15 p.m.: The herd moves into the grass followed by the bull. Only their backs can be seen amidst the tall grass.
>
> 4.00 p.m.: The herd elephants have all gone into the forest. The adult bull and a sub-adult bull (below ten years of age) come back to the lake and bathe.

Elephants may go through an elaborate courtship before mating. The bull may follow the cow, jabbing her with his tusks from behind, they may intertwine trunks, or he may place his trunk and tusks on her back, and so on. When a bull and a cow elephant go into consort they usually mate many times. An actual mating is only rarely seen in elephant populations inhabiting forests, partly because prolonged observation of a courting pair is difficult. To describe a mating scene I have to go forwards several years to a cloudy day in May 1988 at Mudumalai.

The overcast sky was heralding the monsoon. Pre-monsoon showers had already vestured Mother Earth with a green mantle. If spring is the season in the temperate region for creatures to mate and breed, the monsoon is the season to multiply and replenish the tropics.

There were eighteen elephants in all, some bunched together and grazing peacefully in the open glade dotted with young gooseberry *Emblica officinalis* trees, others scattered singly or in small groups amongst the bamboo clumps along the Kalhalla, a rivulet of the Moyar, about a kilometre to the west of Theppakadu in Mudumalai. The gooseberry plants sprouted gregariously from seeds regurgitated by spotted deer which used this glade for bedding at night. The yellowish-green fruit, the size of cherries, had already been plucked by the Kurumbas several months back, put into sacks and carried off to be pickled or used in medicines (gooseberries are very rich in ascorbic acid, or Vitamin C as it is better known).

One young bull and one young cow in that elephant herd seemed to resonate with the spirit of the monsoon. In more prosaic terms, the cow was in oestrus and the bull was mature enough to sense this. The bull was fifteen to twenty years old, while the cow was probably no more than fifteen. She must have been experiencing one of her earliest periods of oestrus. The pair had presumably been courting for some time before I first noticed them. I jotted down the following in my note book.

> 9.05 a.m.: The bull approaches the cow, standing a short distance away from the rest of the herd, whereupon she immediately turns to face away from him. He places his trunk and tusks on her back and after a short while mounts with his hind legs half crouched and front legs astride. Copulation is brief, lasting hardly half a minute. As he dismounts the semen gushes out onto the ground.

The elephant is peculiar among land mammals in that the testes in the male are inside the abdomen. The fully erect penis is flexed upwards near the tip so that it may easily hook on to the vaginal opening which is situated about two feet in front of the cow's hind legs.

The entire sequence occurred so quickly that I barely managed to focus my telephoto lens and take three pictures which, at a slow 1/30 second shutter speed necessitated by the poor light and the 64 ISO film in my camera, proved to be rather unsatisfactory for so momentous an occasion.

9.15 a.m.: The bull follows the cow again for a short distance, jabbing her hindquarters with his tusks. He mounts her again with a semi-erect penis but there is no successful intromission. After dismounting he approaches her again from behind, but she turns to face him, not co-operating this time, and he moves away.

9.20 a.m.: The bull begins to spar with a sub-adult male, placing its trunk over the latter's head in a display of dominance, pushing with locked tusks or simply intertwining trunks. They feed sporadically on grass in between sparring encounters which last less than a minute each. After five or six such bouts they retreat with the rest of the herd into the dense vegetation along the Kalhalla stream bank.

10.15 a.m.: Two females in the herd can be seen feeding on bamboo along the stream. The rest are hidden.

10.20 a.m.: The courting pair emerges out of the thickets and stands by the roadside. The bull mounts the cow soon after this, but since they are facing away from me it is difficult to make out whether they copulated successfully.

10.25 a.m.: Another young couple seems to have caught on to the game. They drift to the roadside besides the first pair in amorous pursuit. The first pair goes up to a large teak tree and starts feeding on small strips of bark.

10.30 a.m.: The entire herd is again out of sight along the stream.

10.55 a.m.: Two elephants, an adult cow and a young bull, can be seen feeding on bamboo. Soon the entire herd emerges and crosses the Ponnangiri road.

The young bull was probably indulging in opportunistic mating in the absence of any large, dominant bull. His courtship repertoire seemed very limited; hence his light-hearted pursuit of the cow and his rather frivolous sparring match at this time with a male much younger to him. Normally the bull would not have been able to mate until he was over twenty-five years old and able to hold his ground against other competing big bulls. But many of the big bulls had gone, falling to the guns of ivory poachers. The younger, sexually mature males now had their opportunity to reproduce and, as I will explain in more detail later, maintain the fertility of the population.

Mating in elephants is not just a matter to be decided solely by the males, based on who is dominant. The female also exercises considerable choice in the matter. At one time it was not fashionable among biologists to talk about female choice in a polygynous mammal. It was thought that the matter was entirely decided by the males. Detailed studies on the Amboseli elephants have shown that the females also decide about which bull to mate with. It would, of course, be advantageous for a cow to choose a bull in musth, if musth was indeed also an indication of the general health and fitness of an animal. A cow would also be wise to choose an older bull, who has already proven his longevity, and thus pass on this trait to her children.

One question people frequently ask me is, 'How do elephants mate? Do they mate only in water?' Many believe that due to their enormous weight elephants need hydraulic buoyancy in order to copulate successfully. Elephants mate on *terra firma* just as other land mammals do (when a bull mounts a cow in shallow water, as it occasionally does, three pairs of feet are firmly grounded!).

The elephant calf is born after a gestation of between nineteen and twenty-two months. Some believe that male embryos undergo a longer gestation than do female ones, but I have not seen any data to conclusively support this claim.

Witnessing the birth of an elephant in the wild would be an extremely rare experience. The closest I have come to observing birth is just after the actual event. Ramesh Kumar, a young biologist from the Bombay Natural History Society and I were driving through Bandipur in the late afternoon of 21 September 1985. At 5.10 p.m. we saw a cow elephant standing not far from the road, inspecting something wriggling amidst the tall grass. We quickly realized that it was a calf. The cow seemed to have dropped it only a short while earlier, perhaps less than ten minutes before we arrived on the scene. There were no other elephants nearby. The cow proceeded to remove the foetal sac with her trunk and then wave it rapidly in the air a few times. She also began throwing mud over her back and on the calf.

The sun had sunk below the canopy, casting its gentle light

on a sacred scene. The jungle was quietly settling down to yet another night, the creatures of the dark having not yet emerged from their hiding places. Mother and child stood frozen for a moment, in homage. We too watched in reverence at this renewal of life.

At 5.45 p.m. the calf made its first sounds, a succession of roars that shattered the stillness of the jungle and announced its arrival into this world. The mother responded immediately, half-crouching and tried to pull her calf up on to its feet, but failed. She resumed her mud bath. At 6.20 p.m. the calf vocalized again. This time the mother began to push the calf forward with her trunk and front feet. After a few such manœuvres the calf rolled down into a small ditch by the road's edge. So far the calf had been well hidden by the tall grass but was now in full view of passing vehicles. The cow stood by her calf in the ditch, feeding sporadically on grass. We left at 7.00 p. m. as it was dark and our presence would be a disturbance to them.

The next morning Ramesh and I were joined by N. Sivaganesan and Ajay Desai, who were also studying elephants under the auspices of the Bombay Natural History Society. When we reached the spot at 7.30 a.m. both mother and baby were still there. This was rather surprising because an elephant calf would normally be up on its feet within two hours of birth. Obviously something was wrong with it. The calf was now roaring frequently, on average once every minute. Its mother kept nudging it occasionally with her trunk or scraping the soil with her feet and throwing mud over herself or reaching out for grass growing at the road's edge. As the sun rose in the sky the calf was exposed to the searing heat. By noon it was vocalizing much less frequently, only once in twenty minutes.

The cow gradually increased her radius of movement, even crossing the road for feeding, but each time she heard the sound of a vehicle she rushed back to the calf. She also began staging mock charges as the vehicles passed by. The highway running through Bandipur and Mudumalai connecting the city of Mysore with Udhagamandalam (or Ootacamund), the well-known hill resort in the Nilgiri hills, is frequented by tourists. That morning a vehicle was passing by every four minutes on the average. The mother elephant was kept busy shuttling between the road and her calf. At times she would stand on the road and the vehicles began piling up on either side. The more adventurous (usually a lorry or a bus) would take the initiative by accelerating forward, with its horn blaring, followed by the smaller cars. This did not always deter the mother. She would rush forward at one of the smaller vehicles, making it veer off the road. We kept in the background, a hundred metres away.

By now it was clear that the calf was in danger of becoming dehydrated. Ranger Belliappa from Bandipur arrived with his staff. A quick discussion was held and it was decided to bring a barrel of water in the hope that the cow might spray the liquid on the calf. Accordingly, a van was sent to fetch the water. The next problem was to deliver the barrel to the elephant. Two men hoisted a wooden pole onto their shoulders, from which the barrel was hung, and bravely walked forwards, followed by the rest of us who were making a lot of noise in order to keep the elephant at bay. The cow would have none of this revelry, however. She charged and as everyone ran back, promptly overturned the barrel, spilling the precious liquid on the road!

The calf was now struggling to get up by frantically kicking its legs. We had a faint hope that it might still survive if given some protection from the blazing sun. The question of taking it into captivity did not arise because it was too young to survive even with good medical aid. There is no known instance of an orphaned wild calf elephant less than two weeks old surviving in captivity even with the assistance of 'allomothers' or domestic cow elephants, and this calf was less than a day old. A young elephant calf cannot tolerate the fat in cow's milk and needs to be given specially constituted milk. It would also need its mother's colustrum during the initial days in order to build up immunity to germs.

The only course of action left was to shift the calf to the shade and to hope for the best. At 1.20 p.m., with a determined noise-making effort, shooting a gun into the air, we finally persuaded the mother to retreat into the jungle. The calf was quickly lifted and transferred into the shade of a tree and hidden from the view of passing vehicles. Water was poured over the calf to cool it. All of us left the scene immediately to allow the mother to return to her calf.

When we returned to the spot at 3.00 p.m. we found the mother by the side of the calf. During the next hour she indulged herself in mud baths and fed peacefully. It was pathetic to hear the calf's feeble sounds. Clearly, it had reached the point of no return.

Most of the vehicles on the highway went past without noticing the elephants. We warned a few who noticed and stopped close to them. Some of them heeded our advice while others did not. The cow was generally peaceful until a car screeched to a halt in front of us. Seven young, fashionably dressed men got out and went up to the elephants. They were clearly in no mood to listen to us. One of them had the bright idea of offering chocolates to the mother. So he went forward, calling out, 'Baby, baby, come and take chocolates.' The matriarch rushed towards the sweet temptation, raising a cloud of dust as she skidded to a halt just short of our jeep. The seven young men fled like jack-rabbits without delivering the chocolates, one of them shooting past the car as the rest piled inside and drove off! Peace reigned again.

We decided to leave soon after sunset. There was nothing more that could be done for the calf. By the next morning the calf was dead. It was a male and had probably arrived prematurely. On average, a calf normally measures about 90 centimetres in height and weighs over 100 kilograms at birth. This calf was only 78 centimetres in height.

It is normal for the adult or even sub-adult cows in a herd to act indirectly as 'mid-wives' during birth. The absence of other elephants at the scene was rather surprising. Perhaps the others in the herd did not stay behind to help because of the disturbance from the vehicles on the road. Selvam described to me an instance when he came across a delivery arena in Mudumalai. The calf had probably been born less than an hour before his arrival. The elephants of the herd were spread loosely in a group around the mother and the new-born calf. The mother had removed the foetal membrane and some of the cows were examining it. Within an hour the calf was up on its feet and had moved away with the herd.

Once the pre-monsoon showers came in May 1981 the elephants left the Araikadavu valley and moved on again, further north. I now saw Meenakshi's family in the open jungles between Punjur and the Suvarnavati reservoir. The herds also spread out into the nearby hills to feed on the freshly sprouting grass. The Karapallam pond was no more an elephant haven. My sightings here reduced dramatically. A few elephants, however, still came here.

I had got through the previous two months without any

untoward incident, even though I was often literally in the midst of elephants. Now for the first time I was given a scare by an angry cow, which served as a warning of what elephants could do. Velayudhan Nair, a forest officer in charge of resource inventories, had come to Hasanur with his team on 7 May. Early in the afternoon we drove up to the Karapallam pond. Thunderclouds were building up in the distant sky and it was steaming hot. When we reached there we saw the back of an elephant that was obviously standing in the depression close to the water. Parking the jeep with Mr Nair in the front seat, I crept up to some bushes along the bund.

I could hear distinctly the flapping of ears as the elephant cooled itself in the torrid heat. Peering cautiously over the bushes, I saw an adult cow standing near the water's edge throwing packs of clayey soil over her back. This would not only help cool her body but also keep blood-sucking parasites away from her skin. After coating her body in this fashion, or by rolling in slush, she would rub herself against a tree trunk and the parasites would fall away along with the dried clay.

I had not earlier seen this cow. A large lump on her forehead made her particularly fierce-looking (later I named her Kali, after a particularly angry Hindu goddess). After a few minutes she abruptly became very still, with her ears spread forwards. A sixth sense warned me that something was wrong. Turning back I saw to my surprise Mr Nair's companions, Ranger Thirugnanam and driver Chelladurai, coolly walking across the open maidan towards the pond though I had told them not to venture near. As I got up, Kali gave a shrill trumpet and came up the slope out of the pond, heading straight towards the jeep, with Mr Nair watching her in horror. I sprinted up to the jeep, started it and pulled away just as Kali skidded to a halt behind it with a second blood-curdling trumpet. Thirugnanam and Chelladurai had bolted off in different directions. Kali, fortunately, decided not to press any further and turned back, but not before having given Mr Nair the fright of his life. The next morning (this time sans Mr Nair) I found Kali again at the pond, accompanied by her three year old son.

The tale of this incident soon spread among the forest department's officials and one of them told Madhav about it. On my next visit to Bangalore, Madhav enquired whether I was taking foolish chances with elephants. He was justifiably concerned because three years earlier a senior forest official, a

The mother comes rushing to sweet temptation.

Divergent Tusks helps himself
to a trunkful of bamboo.

Champaca with her sisters and their children at Kyatedevaragudi.

An entire hillside ablaze at night.

Mr N.R. Nair, had been killed by an elephant while he was out with a researcher in Bandipur.

Actually, I was cautious when it came to dealing with elephants. I was there to study them and to write about them, not to act like a hero in a movie. I mentally planned an escape route before approaching an elephant closely. And provided I took my time and was careful, I found that I could often get pretty close to them. Earlier, the well-known Indian naturalist, M. Krishnan, who had spent several years studying elephants, had warned me, 'When watching elephants, for heaven's sake keep looking behind your back', a piece of advice that came in very handy at times. He had once escaped from a herd in open jungle by smearing himself with elephant dung and crouching behind a bush, while the elephants passed by on either side.

The risks of having to continually interact with elephants were best summed up in a remark by the famous Indian ornithologist, Salim Ali. When Madhav introduced me to him at a meeting and added that I was doing research on elephants, the delightful old man peered at me and enquired in a quivering voice, 'Are you still alive!'

CHAPTER 4

A Megavegetarian Versus the Vegetation

May the leader of the Ganas grant you prosperity, the surface of whose unique frontal lobes is smeared with large masses of the excellent powder vermillion and who with violent movements of his trunk adorns the extensive expanse of the sky and who is engaged in the sport of uprooting multitudes of trees!.

Ratanpur stone inscription (AD 1149–50) of the Kalachuri king, Prithvideva II.

Vinay poked at the *bendai (Kydia calycina)* tree with his left tusk, thrusting it up into the gash and splitting the bark. He grasped a portion with his trunk and tugged expertly with an upward flick, tearing off a four metre long strip. Another tug and the strip broke loose from the tree-trunk and came down. Vinay now began eating the bark, skilfully using his forefeet and trunk to break off small strips before transferring them to his mouth.

After feeding for some ten minutes, Vinay did something that only an elephant can do so effortlessly. He turned towards the tree, and using his forehead and trunk, pushed the tree over. In a minute or so the tree was cleanly uprooted. Vinay tore just one more strip of bark from the tree and then turned away. Almost nonchalantly he began to pluck green grass that sprouted profusely from among burnt clumps. As he wrapped his trunk around a clump and pulled, the tender leaves came off quite easily from their dry bases. Stuffing one trunkful after another into his mouth, Vinay ambled along at a gentle pace.

I had been following him for over two hours along the hill slope to the east of the Hasanur–Dimbam road on 14 June 1981, trying to get to within photographic range. It was just not possible for me to frame him clearly in my camera viewfinder from what I felt was a safe distance, given his reputation. It was quite frustrating. I badly wanted his picture for my file but had to be content with watching him through the foliage with binoculars. After Vinay had moved well away from the fallen tree I went over and measured it; its trunk was 130 centimetres in girth.

To understand why elephants raided crops, I realized that I had to first understand what influenced their feeding in the wild. In the literature I found widely varying opinions as to whether elephants were mainly browsers, feeding on shrubs and trees, or grazers who feed on grass. Many researchers have observed elephants in mainly grassland habitat, such as the African savannas or the swampy *villus* in Sri Lanka, where they can be easily seen. Elephants would anyway eat mostly grasses here. Since elephants also spend considerable time in browse-rich forest, either on a daily or a seasonal basis, it is also important to observe them in this environment and to estimate the proportion of time which they spent there.

During the first two months, the dry months of March and April, I observed that the elephants browsed mainly on the leaves and bark of a variety of shrubs and trees. The Araikadavu valley, with its dense thickets of shrub acacias, was a favourite feeding ground for the herds. The thorny shoots of the acacias did not deter them in the least. They would break off the shoots, and then pack them together with their trunks, before transferring them to their mouths and grinding them up with their large molar teeth. Near Karapallam I noticed that a gregarious stand of the pretty cream-colour barked *Acacia suma* tree had been attacked, stripped of its bark and broken by the elephants. Along the stream bank they would pull at the stunted culms of bamboo, consuming the leaves and branches alike. In the nearby hill slopes the elephants fed little on the tall grasses growing there but preferred the bark of trees or the underground root of the shrubby palm *Phoenix humilis*. When I went north to the moist deciduous forests of the Biligirirangans, I saw that elephants had stripped the bark of many malvaceous and related plants, including the trees *Grewia tiliaefolia* and *Kydia calycina,* and the shrub *Helicteres isora*.

Now with the onset of the monsoon in June the herds had moved away from the Araikadavu valley. As they ranged over the hill slopes they avidly fed on the new flush of tender, tall grasses, *Themeda cymbaria, Themeda triandra* and *Cymbopogon flexuosus*. They seemed to be especially fond of the grasses that sprung up afresh in areas that had earlier been burnt by fire. Whenever they moved into the short-grass areas they did feed on the shrubs and trees, and even in the tall-grass forests they supplemented their diet with bark, but there was a distinct preference for grasses during the early wet season.

As the months rolled by and the grasses grew taller and coarser, the elephants increasingly avoided eating the top leaves but instead consumed the basal portion. They would uproot clumps of the tall grasses, skilfully clean the roots of mud by thrashing against their forefeet, chew off the succulent basal portion and discard the fibrous leaves. In the open jungles the short grasses grew with the coming of the rains and by October or so the elephants could pluck sufficient amounts of them.

The two bamboos growing there, the taller, moisture-loving *Bambusa arundinacea* and the shorter *Dendrocalamus strictus*, were favourites almost all year round. The former grows mainly along the banks of streams, while the latter covers the drier hill slopes. The bamboos flower only once—at the end of their life span—and then die. *Bambusa arundinacea* flowers gregariously once every forty-five years. This means that the entire bamboo population over a vast region flowers at the same time, producing vast quantities of seed, only to perish after this 'big bang' reproduction. The last gregarious flowering in the Nilgiris and Eastern Ghats was during 1972–3, which means that the next major flowering can be expected only after 2017 (although I did see mass flowering in Mudumalai during 1990–1).

Botanically the bamboos are grasses (often called 'tree grasses') but the manner in which elephants feed on them is more akin to browsing. Elephants consume not only the lateral shoots bearing the leaves but also the tough culms, pulling them down with their trunks and breaking them under foot. When an elephant herd parks itself in a bamboo grove for feeding, the jungle resounds with the sharp, rifle-shot like cracking of the slender culms.

In addition to their staple foods—grasses, bamboos, legumes and the bark of selected plants—elephants have an amazing variety in their menu. Various climbers, creepers, palms and succulents are eaten. The leaves of various species of fig trees are much sought after. Fruits such as that of the tamarind *Tamarindus indicus*, the wood apple *Feronia elephantum*, the wild mango *Mangifera indica* and *Careya arborea* are seasonal delicacies. By their fruit-eating habits, elephants aid in the dispersal and subsequent germination of seeds. The passage through the elephant's gut helps break the dormancy of seeds and speeds up their germination.

In recent years, elephants have taken to eating the bark of the exotic *Eucalyptus*. Although the eucalypts were introduced into the country from Australia over a century ago, there is no mention in earlier writings of them being consumed by elephants. A herbivore obviously has to approach a new food plant with caution because it may contain toxins its systems cannot neutralize effectively. It may take a number of generations of sporadic feeding, imitation by others in the population and the development of tolerance before a new food is embraced by the entire population. The bark of teak, for instance, is frequently consumed in the south but very rarely in the northern part of the country, where it has been planted in recent decades.

To imagine that an elephant uses only brute force to obtain its food is highly fallacious. Picking up a tiny mimosa flower is as facile as pushing over a large tree. No other land animal is able to exploit such a wide range of plant resources. This is made possible by that unique organ, the trunk, whose finger at the tip can delicately pick up a tiny object and whose reach extends to a tree five metres high.

My final list included 112 plant species consumed by elephants in my study area. This is not unusual for an elephant population. Robert Olivier listed over 400 plants, mainly palms, that were potential food for elephants in the rain forests of Malaysia, while N. Ishwaran observed the Sri Lankan elephants consume over 100 species.

Animals obviously have their own likes and dislikes among potential food plants. This may be influenced by their dietary requirements and the nutritive values of potential foods. Analyzing the food plants for their nutrients, I saw that the changing protein content in plants was one factor that influenced dietary choice. The new flush of tall grass that emerged with the rains had about ten per cent protein, which was more than adequate for the elephants to maintain good body condition (they need at least five per cent food protein for this). As the grasses grew and became increasingly siliceous and unpalatable, the protein value dropped gradually, especially in the succulent basal portion, until it was no longer sufficient for the elephants to be able to maintain their condition. Many of the shrubs and trees, chiefly the legumes, retain ten to twenty per cent protein in their leaves even during the dry season, and hence were clearly preferred to the grasses at this time.

Many explanations have been given for elephants eating bark. Since bark is very fibrous, the roughage taken along with fresh foliage may help maintain the correct protein to fibre ratio for proper digestion and also prevent constipation. Bark is also rich in minerals compared to other foods such as grasses.

Elephants need a lot of calcium, not only for the growth of their bones but also for their tusks (which are 45 per cent calcium). Pregnant and lactating cows will also have a special need for this mineral. Sodium is another mineral that elephants will go to great lengths to obtain. They commonly eat soil or drink from ponds that are rich in sodium.

Putting together the data on seasonal feeding preferences in different habitats and the time they spent in them, I calculate that elephants feed more on grass (70–75 per cent) during the early wet season, on roughly equal quantities of grass and browse during the late wet season, and more of browse (70 per cent) during the dry season. My direct observations thus indicate that elephants on average consume grass and browse equally during the year. I wanted to verify this in some other manner and the means to do this fell into my lap quite accidentally.

Back at the institute in Bangalore, I was thumbing through an issue of *Nature*, the scientific journal, when I came across an article on the origin of maize cultivation in North America which had been deduced from the ratios of two isotopes of the element carbon in fossil human bones. It struck me that I could use this technique to find out in what proportion elephants consumed grass and browse.

The principle behind the method was fairly simple. The ratios of the stable carbon isotopes, the rare 13-carbon and the common 12-carbon, were different in shrubs and trees compared to grasses (with the exception of bamboos). These 'isotopic signatures' would be preserved in the tissues of animals feeding on the different types. For analysis, the protein collagen which is found in bone is usually the most suitable.

I began collecting the bones, usually the lower jaws, of dead elephants for this work. I still needed a mass spectrometer to determine the carbon isotope ratios. Fortunately, I located one at the Physical Research Laboratory in Ahmedabad. The scientists there, S.K. Bhattacharya, R.V. Krishnamurthy and R. Ramesh, were using the instrument to study the past climate, but were happy to collaborate in my elephantine venture.

The results from the isotope analysis of bone collagen revealed something interesting. In adult elephants between 55 per cent and 83 per cent of the carbon in the collagen came from browse plants. This clearly meant that the browse plants were contributing proportionately more protein to the elephant's growth than did the grasses, even though the two food types were equally eaten. Clearly, then, the browse plants were very important for the elephant's nutrition even though it is well adapted to grazing.

Since the Oligocene period (25 to 40 million years before present) the ancestors of present-day ungulates, including elephants, have adapted in a number of ways to deal with the vast reservoir of low-quality compounds, such as cellulose, that make up the plants. The vegetation has also changed in response to a cooler and drier climate. One way for the animals to cope with the poorer quality of the food was to eat more of it. They thus increased in body size. The teeth took on complex surface patterns in order to grind the increasingly tough and fibrous plant material, and their crown height increased, prolonging the life of the teeth and their bearers (the animal). The complex modifications in teeth structure are best exemplified in the evolution of the elephant.

Herbivorous mammals do not produce the chemical enzymes needed to digest cellulose, but many microbes do. Ruminants such as cattle and deer have evolved a fermentation chamber, the rumen, in which microbes do the job of breaking down the cellulose for them. Non-ruminants, such as elephants, achieve the same-thing in a less efficient manner. Their caecum and colon are huge fermentation vats in which bacteria and protozoa break down cellulose.

An infant is not born with the microbes necessary to aid cellulose digestion; it has to acquire them from its surroundings. Among ruminants the calf acquires the microbes during mutual licking with its mother or by eating plants which have been coated with rumen fluid by older animals when they chew the cud. Young elephants acquire the necessary organisms somewhat differently—by eating the dung of older animals, a habit known as coprophagy.

On two occasions I observed young wild elephants sampling the dung of adults. One was a two year old male which ate small amounts of its mother's dung while she was grazing nearby. Another time, while observing a family of seven elephants I saw an infant, hardly a couple of weeks old, touching fresh dung with its trunk and keeping the tip in its mouth. No quantity of dung was taken in but a sample of microbes would undoubtedly have been transferred.

The digestive efficiency of a large herbivore is relatively low. More than half a century ago a classic experiment by Benedict on a captive Asian female elephant named Jap showed that only 44 per cent of food protein is digested. More recent experiments on captive African elephants have shown that the digestive efficiency may be as low as 22 per cent.

An elephant eats six to eight per cent of its body weight as fresh food every day. Thus a full-grown bull weighing four tons would daily consume anywhere between 240 and 320 kilograms of forage. Three-quarters of this would be moisture; thus 60 to 80 kilograms of dry matter is taken in. An 'average' elephant would, of course, eat much less food, about 100 to 140 kilograms of fresh forage daily. My observations showed that elephants could meet this requirement by feeding for twelve hours in the day, but that they could manage this intake in only six to seven hours feeding in cultivated fields!

It is easy to see that with such prodigious requirements elephants can cause drastic changes to the vegetation when there are too many of them in relation to the supporting vegetation. Exactly how many is too many has never been satisfactorily resolved. Nevertheless, when a forest begins to disappear and grassland takes over and elephants damage or push over trees, there is cause for concern.

Behaviourists have speculated that tree-pushing by a bull could be a form of social display, a message to other bulls in the vicinity about who is stronger. The day in June 1981 when Vinay pushed over a large *Kydia calycina* tree on the slopes near Dimbam, another younger bull called Cradle Tusks was moving in his company. The two remained together without any apparent antagonism for nearly a week, but then neither of them was in musth.

Harvey Croze, working in the Serengeti, found that tree-pushing is more likely to be merely for obtaining nutrition rather than for social reasons. Very often it is a younger bull in a bull group that first pushes over a tree, only to be supplanted while feeding by a more dominant bull.

Whatever the reasons for pushing over trees, such behaviour has often led people to label the elephant as a wasteful feeder. From a neutral biological perspective this value judgement does not seem justified. True, a grazing deer utilises the plant resources more efficiently than does an elephant which knocks down a tree for a mouthful of fodder, or even, seemingly, for the fun of it. However, the loss of a tree in an elephant-dominated landscape may be part of the normal biological process. The fallen tree may provide critical food for smaller animals and a micro-habitat for a host of creatures. The plant saplings in the vicinity respond with a new-found vigour to the gap created in the cover. In any case, elephants are no more wasteful feeders than are humans in energy-intensive technological societies. And elephants do not need non-renewable resources to package their food!

Elephants do make a tremendous impact on the vegetation, however. During the past two decades there has been a lively and often acrimonious debate about whether such changes in vegetation caused by the African elephant, from woodland to grassland in some regions, is desirable or not. This in turn has generated conflicting and often emotional views on culling of elephants.

At one extreme there are those who consider all such change as unnatural and as being caused by the compression of elephants into smaller and smaller areas and higher densities because of an expanding human population. It has even been suggested that elephants could have been responsible for creating deserts in parts of Africa. According to the proponents of this view, there is no alternative to culling elephants (a euphemism for killing them in the African context) in order to maintain their populations below the 'carrying capacity' of the habitat.

Elephants have been or are being culled as part of official policy in Uganda, Zambia, Zimbabwe and South Africa, among other countries. The culling is usually carried out in accordance with a management plan. Entire families are shot at the same time. The meat is often processed for consumption, the skin is tanned into leather, while the ivory is removed and sold. Considerable revenue is generated for the local economy. The operation may be combined with a research programme. The post-mortem examination of large numbers of elephants has yielded valuable information on their reproductive biology, physiology and population dynamics, which would have been impossible to obtain otherwise.

Others have taken an entirely different view of the 'problem'. The conversion of woodland into grassland is here seen as a natural part of long-term ecological processes, and in many instances elephants are being wrongly blamed for the disappearance of trees. David Western and C. van Praet found, for instance, that the fever tree *Acacia xanthophloea* in Amboseli, Kenya, was being killed by a rising water-table which was causing increased salinity of the soil, and not by elephants. Harvey Croze showed that fire, and not the elephant, was the chief culprit in preventing adequate regeneration of *Acacia tortilis* in the Seronera region of the Serengeti.

Graeme Caughley, a population ecologist now based in Australia, suggested that elephants and trees may never exist in equilibrium in East Africa. He proposed a model in which elephants and trees are locked into a cyclic relationship, where elephants increase in numbers and thin out the woodlands, only to decline later, allowing the woodland to recuperate. He further suggested that the period between successive peaks or troughs in the population density of elephants or trees was about 200 years in the Luangwa Valley of Zambia. The non-equilibrium nature of living systems has been demonstrated for a wide range of communities, from coral reefs to rain forests. Since the human lifespan falls far short of such time periods, it is easy to realize why we so poorly understand such natural phenomena and instead engage in fierce, acrimonious debates.

I extended Caughley's basic model to cover elephant–vegetation interaction over the entire range of habitats from equatorial rain forest to semi-arid savannah woodland. Across this diversity of habitats in which elephants are found, there are clearly tremendous differences in the manner in which they interact with trees. No one has suggested that elephants are destroying evergreen rain forests. In the tropics, not only does it rain heavily in these forests, it also rains more consistently year after year when compared with the semi-arid savannahs where the inter-annual variation in rainfall is much greater. The more predictable rain forest habitat is a relatively stable community as opposed to the fluctuating savannah habitat. The huge trees in the species-rich rain forests cannot be pushed over by elephants. Even the plants in the understorey are often not palatable as they contain many defensive toxic chemicals. The rain forests thus support only low numbers of elephants, usually less than one elephant for every ten square kilometres of area. Elephants and rain forests exist in near stable equilibrium.

The very opposite is true of the semi-arid savannahs, which are low in tree species diversity. Most of the woody plants here are potential food for elephants. In addition, the abundant grass production increases the carrying capacity, such that anywhere from one to five elephants may be found for each square kilometre of habitat. The destruction of woodland is very noticeable when the elephant population reaches a peak. Elephant and tree populations undergo cyclic increases and declines. In between these extremes there exists a range of intermediate habitats with corresponding gradations in the elephant–vegetation cycle.

For such a scenario to work there must be proximate mechanisms driving the increase or decrease in elephant numbers. One possibility is that fertility may decline in elephant populations which are forced to increasingly feed on protein-deficient grasses as woody plants disappear. Richard Laws and his colleagues, who worked in Uganda during the late 1960s, have shown that this can happen to African elephant populations. A long-lived animal such as the elephant is naturally resistant, by virtue of its low death rates, to sudden changes in its demography. By the time a decrease in fertility leads to a negative population growth rate, the population greatly overshoots the environment's carrying capacity. There is also some historical evidence that elephant population densities undergo considerable changes over time scales of decades or centuries. At the beginning of this century the Tsavo region of Kenya had far fewer elephants than in recent times. G. P. Sanderson observed in a lecture (reported in R. A. Sterndale's *The Natural History of the Mammalia of India*) that there were hardly any elephants in the Biligirirangans towards the end of the eighteenth century, but had increased from then onwards. It is difficult to conjecture whether this was due to natural population cycles or changes in human pressures on the habitat.

Some data I gathered on elephant damage to trees indicated that in the area in which I was studying there was no immediate cause for alarm. My studies were preliminary and looked at only four trees, *Kydia calycina, Grewia tiliaefolia, Acacia leucophloea* and *Acacia suma*. Although up to 80 per cent of *Kydia* and *Grewia* plants were utilized by the elephants, less than 15 per cent of the standing trees had died, and then this was not always due to the elephants. Broken trees regenerate quite well; if the entire tree is pushed over by an elephant or killed by fire the plant regenerates through root coppices. *Acacia leucophloea* seems to undergo population cycles in relation to environmental factors, similar to those of *Acacia xanthophloea* in the Amboseli.

Near Karapallam *Acacia suma* seemed to be clearly declining due to elephants' damage during 1981–2. Over 40 per cent of stems in some size-classes were dead. But during subsequent years there was little damage, and in 1987 the trees were very much still there, though perhaps a little thinned out. Strangely, there were no saplings, even though these were present in 1982. Seed germination and sapling growth seem inhibited by factors other than elephants. This clearly meant that plant populations have to be studied over long time periods before we make any definitive statements on their dynamics, a project that I later embarked upon.

Even if it is clear that the tree cover is declining due to elephants, what then should be the course of action? This would depend on the conditions peculiar to a region and the defined management objectives. It may be essential to preserve a particular tree community for its aesthetic value or because it is rare. Endangered animals may be genuinely threatened by habitat changes. The destruction of woodland may increasingly force a large elephant population, confined within a small area, to seek more nutritious forage from crop fields. In such cases, some culling of elephants may be justified.

On the other hand, one might decide to let nature take her own course and to consider the disappearance of an entire woodland or a particular tree species as part of a cyclic relationship between elephants and vegetation. If a sufficiently large habitat area is available, the localized shocks delivered by elephants to tree stands may be easily absorbed. Inter-annual differences in the elephant's movement patterns or changes in preference for a plant in relation to its declining abundance may mitigate the problem over time.

No one has yet seriously suggested that any Asian elephant population is destroying its habitat and needs to be culled, though this may arise in the future. With the halting of the capture of elephants and the continuing loss of habitat in many regions of India, the compressed elephant populations may increase to very high densities. Added to this, the construction of dams, both small and large, and the digging of ponds as part of reserve management, leads to the maintenance of elephant populations at artificially high levels by providing perennial water sources even during the dry season. In any case, the cultural traditions in Asia largely rule out the outright killing of elephants. The only alternative, if such a need arises, is to capture elephants for domestication.

One afternoon in June 1981 I received a message from V. Krishnamurthy, a veterinary doctor in Tamil Nadu's forest department, asking me to come over to Mudumalai for the annual measuring of the captive elephants. This was a good opportunity for me to not only get data on the growth of elephants but to also observe them. Dr. K, as he is commonly and affectionately known, is one of the most experienced elephant doctors in the world. He had carefully preserved old records of elephants captured or born in captivity since the late nineteenth century and was only too happy to let me delve into his registers. We took a fresh set of measurements for all the elephants. In addition to height, we measured the circumference of the front feet and the girth of the tusks at the lip line. These elephants had been weighed annually since 1976 in a weigh-bridge meant for lorries.

With all this data I was able to work out the relationships between shoulder height and age, weight and age, tusk girth/weight and age, shoulder height and the circumference of the front feet and so on. Each of these relationships proved useful at some stage or the other in my work. The weight–age relationship enabled me to estimate the biomass of the elephant population, the circumference of the front feet came in handy when I had to age crop raiding bulls, while the tusk–age relationship was used to calculate the quantities of ivory poached. Most important, I was able to use the shoulder height–age relationship to assess the age of wild elephants. There were some problems in applying these results directly to wild elephants, however. The captive elephants were clearly stunted in growth when compared to wild elephants. However, I was able to correct the results before applying them to wild elephants. These results were published in collaboration with my colleague Niranjan Joshi and Dr. K in the *Proceedings of the Indian Academy of Sciences*.

When working at the Mudumalai elephant camp I also carried out trials with captive elephants of the photographic method of measuring shoulder heights. I was pleased to find that the error was no greater than about four per cent, with an average error of only two per cent. This was certainly an acceptable level of accuracy.

The visit to Mudumalai gave me a chance to observe wild elephants and to compare their feeding habits with those in the Biligirirangans. With the monsoon having set in, the forest was a lush green and the elephants had moved into the tall grass deciduous forest around Theppakadu. Here, too, they fed mostly on grass during this season. During the short visits I made later to Mudumalai I noticed that the broad patterns of elephant movement and feeding were similar to that of the elephants in my main study area.

Once again Selvam and I acquainted ourselves with the large bull we had seen the previous November. Divergent Tusks was a familiar figure in Mudumalai and Bandipur during the wet season. From June to August he would invariably haunt Kargudi. Vehicles would tail back along the Mysore to Ooty highway as Divergent Tusks systematically dealt with the tall bamboo fronds along the road, stubbornly refusing to pay any heed to the blare of horns. One evening he climbed out of the Moyar and coolly walked across the makeshift football field at Kargudi while a game was in progress.

For Selvam, the nights were filled with the breaking of bamboo culms behind his tiny room on the banks of the Moyar. From his window he could put out his hand and actually slap Divergent Tusks on his back, if he so wished. One night, we retired after midnight after watching him feed on a teak sapling. The next morning we found his footprints just outside the flimsy front door. Yet with Divergent Tusks there was no cause for worry. He was as gentle as a bull elephant would ever get.

Siddharth Buch and Jagannatha Rao also came to Mudumalai at this time and launched themselves enthusiastically into measuring elephants and photographing them. All of us had a glorious time going around the jungle watching Mudumalai's big, magnificent tuskers. They seemed to be especially fond of parking themselves close to the road at night to feed on the luxuriant bamboo growing along the Moyar. Every night we would invariably brush past them as they stood on the sharp bends of the road. If one of them had turned around suddenly in panic we could have collided.

After two weeks of Mudumalai it was time for me to get back to Hasanur. When I made the rounds of the study villages I found the farmers in some of them waiting with complaints of crop-raiding elephants. My time was taken up in trudging through millet fields, following in the footsteps of the raiders, measuring the extent of the damage, weighing plants, and carrying out the less-glamourous part of the work.

CHAPTER 5

Raiders and Rogues

> Oh, Lord of the Kurinji land, where the Kuravas raised a big cry with arrows, drums and slings in their hands and whistles in their mouths on seeing the approach of the elephant with fiery eyes, alone and separated from its mate.
>
> (This is) the place where the leader of the herd separates from its kith and kin and waits in ambush by the roadside to get at the passers by.
>
> Passages from Tamil Sangam Literature (1st to 5th century AD).

It was 3 November 1981. The air was distinctly nippy and humid as the result of a shower earlier in the day. Twilight was fast receding into darkness as Setty and I made our way through the *ragi* fields of Hasanur armed with flashlights. The *ragi* had flowered in most of the fields and Vinay had been making nocturnal raids on the crop during the past month. The previous three nights he had been entering from the northern boundary. In most of the fields the farmers and their watchmen had already taken up their positions, huddled around a fire inside their flimsy thatched structures which were built either on the ground or up a tree. Beams of light criss-crossed the darkness, punctuated occasionally by the full throated shouts of the men communicating and reassuring one another from their respective fields.

In one field our lights suddenly picked up two gleaming objects—a pair of tusks. A cold shiver went up my spine on seeing a bull so close to. However, the bull, which no one else seemed to have noticed, quickly retreated.

Around 7.00 p.m. we were informed by a Sholaga family that a bull was feeding in the field of H.S. Basavappan. As we approached his field we could see lights and hear shouts. When we reached the spot we were informed that the bull had entered the adjacent field about half an hour before and was refusing to leave. I climbed up into a machan in Basavappan's field, from where I could make out an elephant feeding calmly in spite of the men's attempts to chase it away. At exactly 9.00 p.m. the elephant moved into Basavappan's field where it continued feeding on the *ragi* plants about seventy-five metres from our tree. I could now recognize it as Vinay; only the left tusk was reflected in our flash lights.

After the repeated attempts to dislodge Vinay proved futile, Basavappan resigned himself to his fate and went to sleep. I huddled close to the fire which was kept going throughout the night, watching Vinay from the safety of the platform which was well beyond his reach. In the past he had stuck his left tusk through the wall of a hut, attempted to pull down the thatched roof of another, and given the inmates of a third a scare by charging their flimsy dwelling. Tonight he was strangely calm, apparently relishing the tasty flowering shoots of the *ragi*. The silence of the night was broken only by the gentle crackling of the fire and by Vinay pulling out clumps of the succulent plants. I snatched a few bouts of sleep in between keeping a watch on Vinay. He fed continuously until the first hint of dawn at 5.50 a.m. He then stopped abruptly, put his trunk up in the air for a moment and quickly made his way back northwards to the jungle some half a kilometre away. He was now barrel-shaped, engorged with a quarter ton of fresh *ragi* plants.

Variations of the same theme, crop raiding, were enacted some 120 nights in the year by Vinay. He was easily recognizable even at night by his broken right tusk. After I had moved into Hasanur in late February 1981, the first complaint about him came from Talamalai. He had trampled a *ragi* field and pushed over some coconut trees. By May he had shifted his attention to the maize and sorghum crops at Hasanur. It was towards the end of September that he really came into his true element, when the *ragi* crop began flowering in the fields of Hasanur. From then onwards he raided crops every night until the harvest early in January.

This pattern is largely true of all crop raiding by elephants. After the *ragi* harvest in January, the fields are fallow until the onset of pre-monsoon showers in late April. During these dry months only the bulls occasionally venture into the fields, usually in search of perennial crops such as coconut, banana, mango and jack fruit. Once the showers come, some farmers cultivate maize or sorghum during the first wet season from May to August. During this time both elephant bulls and family groups indulge in raiding these crops.

It was at this time of year that I came to know of another notorious bull which inhabited the western portion of the study area. Here, a small patch of forest known as the Akkurjorai Reserve had been completely encircled by cultivation and separated from the main Talamalai Reserve Forest to its east. This did not, however, deter an exceptionally large bull from walking at will through the cultivated tract between these two reserves. Its front foot circumference measured from clear imprints in the fields was about 155 centimetres; the bull would have been over three metres in height.

On the afternoon of 7 June Setty hurried up to me with the news that a man had been killed by an elephant in a village called Mallanguli. We rushed there to inquire into the incident. When we reached the village a crowd soon gathered around us to narrate what had happened. The previous night, two brothers had been keeping watch on their sorghum crop from a thatched hut when an elephant came into their field. They had shone a light to look at it; the elephant reacted violently, smashing their hut and killing one of the men. When we inspected the field we found clear signs pointing to a very large bull. We met the surviving brother, who was still rather shaken but who seemed relieved that he had not been the victim.

The same bull stuck again the next month. In the early hours of 31 July it killed an old woman who was sweeping the ground in front of her hut at Bokkapuram village. During the day it then went on the rampage in the nearby maize and sugar cane fields, drank a barrel of toddy and by evening entered the town of Talavadi. Here it began pushing at the mud-walled houses with its tusks. A woman and her baby had a narrow escape when the wall and roof of their house collapsed. The bull was not yet satisfied; it then disposed of a goat and overturned a bullock cart, losing a piece of one of its tusks in the process, and then injured a man before heading towards the Suvarnavati Reservoir.

After Setty and I received news of this two days later we went to inspect the damage and found the injured man, Mahadeva, in the local hospital. He complained of pain in the ribs and wanted to be taken to a larger hospital at Satyamangalam. Loading him and his wife into my jeep, we made the two hour journey to Satyamangalam that night, with Mahadeva groaning continually in the back and his wife berating him for exaggerating the extent of his pain. She said that the elephant may not have actually hit him but that he may have fallen into a ditch out of fear, for how else could he have survived the bull.

I regularly received complaints about this bull from other farmers in the tract. It could be easily identified by its exceptionally large footprint size. This bull seemed to be the only elephant using the Akkurjorai Reserve regularly; hence, I simply called it the Akkurjorai Bull. Its whole movement pattern was geared to feeding on cultivated crops in the tract between the Akkurjorai and the Talamalai Reserves. This also seemed true of many other bull elephants which lurked in the vicinity of cultivation for many weeks at a stretch when the crops were full-grown.

On the other hand, the family herds seemed to raid crops largely when they encountered these in the course of their natural seasonal movements. By June, Meenakshi's clan had moved away from the Araikadavu valley into the open jungle to the north. At this time many herds were reported entering the villages of Punjur, Kolipalya and Gumtapuram to feed on the maize and sorghum crops being grown here. Some of these raids inflicted considerable damage. On the night of 5 July a herd of about twenty elephants trampled three hectares of maize fields at Mudahalli, a small hamlet near Kolipalya. For the farmers who collectively cultivated the fields here it was a severe blow; they lost half their crop that single night. I was never sure which elephants were involved in these raids, but it was not unreasonable to assume that members of Meenakshi's clan were responsible.

The connection between the seasonal movements of elephant herds and crop raiding became even clearer when *ragi*, the staple food crop of the people in the area, was being cultivated during the second wet season. Beginning late September, the flowering *ragi* plants had a magnetic attraction for elephants, both for individual bulls and herds. Champaca's clan had come south from the dense moist deciduous forests in the hills to the open jungles near the villages of Punjur, Kolipalya and Gumtapuram. The herds now began regularly raiding the crops there. Meenakshi's clan may also have indulged in some raiding, though I am not certain of this. At this time they were moving further south, sequentially raiding Chikkahalli, Neydalapuram, Talamalai, and Mavanattam.

Once raiding began during the *ragi* cultivation season, most of the large villages suffered damage almost every night from either adult bulls or from herds, or from both on some nights. The adult bulls took up position near villages and raided crops regularly for anywhere between one and three months at a stretch. Up to six bulls came into Hasanur on certain nights, four bulls in two pairs concentrated on Punjur and Kolipàlya, while three bulls, including a tuskless one, operated at Chikkahalli and later at Neydalapuram. I tried to get acquainted with at least the bulls that came into Hasanur. This was easier said than done because at night it was difficult to identify the bulls which had both tusks intact unless one got fairly close to them with a powerful flashlight.

One large bull, which I never got to name, reacted aggressively towards people who made a noise or shone a light at it in the fields. On the night of 27 October two men barely managed to clamber up a tree, one of them emptying his bladder out of shock, before the bull reached the thatched shelter beneath the tree and smashed it. This bull was later shot with a muzzle-loaded gun when I was away from Hasanur. It managed to walk over a kilometre through the fields before it collapsed inside the jungle. When I returned to Hasanur some days later I heard rumours that an elephant's carcass was lying somewhere nearby. Nobody was willing to say exactly where. The reason soon became obvious. When the carcass was finally located the tusks were missing. Apparently the men who were chased up a tree had shot the bull a couple of nights later, but since the incident went unnoticed they removed the tusks.

During my night patrols at Hasanur I encountered two more bulls which several times crossed the road in front of me which ran along the forest boundary. One was the Karapallam Bull, with a third of its left tusk broken, while the other was a younger bull I could not identify. That same evening in October, a few hours before the two men had almost fallen victim to the resentful bull which they later killed, I wrote the following account in my book:

> While driving back from Dimbam to Hasanur just after 7.00 p.m. a medium-sized bull with part of its left tusk broken (the Karapallam Bull?) and a smaller bull were coming from the west to cross the road. As soon as I had passed them they quickly crossed over to the fallow land on the east. I turned my spotlight

> on them. The larger bull turned towards me with its trunk raised, curling it around its broken left tusk, seemingly in nervousness. It took a few steps forward as if to charge, but retreated, moving parallel to my jeep. The younger bull kept in the background, searching for a place to get down into the Araikadavu. Within ten minutes the two bulls disappeared into the stream and presumably went over into the *ragi* fields on the other side.

This kind of association between bulls for raiding crops is common. A solitary existence is the norm otherwise; 93 per cent of all my sightings of elephant bulls inside the forest during the day were of solitary creatures. There were a few sightings of pairs and just one instance of three bulls associating together. On the other hand, only 57 per cent of bulls involved in raiding crops were solitary, while the rest were in groups of two, three or even four.

Before drawing any definite conclusions about the significance of this behaviour, I had to be sure that such groupings were not accidental ones arising simply due to the fact that several solitary bulls independently had converged upon the same field (which may have had a particularly attractive crop). Examining the boundary of the cultivation with the forest, I found that in most cases the bulls had indeed entered together, as deduced from their footprints or by questioning farmers who lived along the periphery. Since these cultivated tracts had boundary perimeters of ten to fifteen kilometres and the bulls could have entered at virtually any point, it was unlikely that they had all converged upon the same entry point merely by chance. The fact that there are associations among bulls for crop raiding was thus established. The bulls involved seemed to go their own way inside the jungle during the day, perhaps maintaining contact with one another, and to join together in the evening before entering the cultivation. Some of these associations lasted for up to a month at a time.

What could be the advantage of teaming up with other bulls while raiding crops? One obvious reason could be a higher success in raiding as a group than when solitarily. The grouping could enable bulls to tackle hostile farmers more effectively. The tendency for bulls to associate in larger groups while raiding was significant only during the *ragi* cultivation season, when crops were available in plenty and it would be more advantageous to co-operate, rather than to compete, in obtaining this resource.

Another interesting pattern emerged from looking at the ages of the associating bulls versus those of the solitary raiders. I measured the circumference of the front feet of the raiding bulls from the clear imprints they left in the fields and aged them from these measurements. It turned out that the solitary bulls raiding crops were mostly above twenty-five years old. In the few instances when younger bulls ventured alone into the fields they were chased away relatively easily. On the other hand, the raiding bull groups consisted of only big bulls, or at least one big bull accompanied by smaller bulls. It thus seemed that the relatively inexperienced young bulls needed the company of an older bull in order to raid successfully, while the experienced bulls such as Vinay and the Akkurjorai Bull could manage on their own. Perhaps the small bull tagging along with the Karapallam Bull simply needed the security of the latter's presence or to learn the tricks of raiding.

Apart from the fact that a lone cow elephant with a calf will hardly take the risk of entering cultivation, the family herds did not show any special tendency to coalesce into larger groups for raiding. This is perhaps understandable because the herds already have sufficient numbers for defence and to further increase their size would only reduce the amount of forage that each individual would get when feeding in closely-spaced groups.

By late November 1981 many of these patterns were already clear enough to me, even though I had not yet formally put all my data together. The male elephant was the more inclined to raiding, and its life seemed to revolve around the tempting cultivated fields. The female herds, on the other hand, seemed to generally come into cultivation only when they encountered these during their seasonal migrations. The peculiar pattern I observed at Hasanur strengthened these suspicions. This village remained free of raids by herds until the end of November, although the *ragi* crop had been at the stage most preferred by the elephants since late September, simply because their normal movement during these months did not bring them close to this village. It was only on 30 November that a herd of seventeen elephants moved south through the Araikadavu valley and began raiding a farm at Binakanhalli, an offshoot of Hasanur located behind my bungalow. On the evening of 4 December I went to the farm and made these observations.

> It was quite dark at 6.30 p.m. when a herd of seven elephants appeared at the northern boundary of the farm. There were three sub-adult males and the rest were cows of different sizes. This was a sub-group of the larger herd that had moved into this area recently. The ripened *ragi* grain had been plucked only a few days earlier leaving only the stalks. The elephants began feeding on the *ragi* stalks. A battery of lights from my jeep and the farm's tractor did not perturb them. Fire crackers were burst twice but again they did not react. Balakrishnan Naicker, the farm owner, fired twice above them with a light rifle but again, surprisingly, they did not seem to be disturbed and continued feeding. Finally, he fired at a sub-adult male. With a scream it ran back into the jungle, followed quickly by the rest of the herd. When Setty and I were returning we saw the remaining ten elephants just inside the jungle but they retreated immediately upon seeing our lights.

The next night, the elephants again came to the farm, though I did not see them as I was engaged in other work. On the morning of 6 December, Setty and I picked up the trail of the herd from the farm and followed it up the hills to the west. Perhaps they were going up to Hulikere ('tiger pond'), an hour's walk for a reasonably fit person. Near Hulikere there was a large cattle *patti*, but the elephants nevertheless used the pond for drinking and bathing. Two kilometres south of this was another large tank, Alamalaikere, which had a nearby small temple to which people came for worship on Mondays. After raiding the fields of Hasanur, Vinay would sometimes head towards this tank.

We soon inferred that the herd had split into two groups since we could hear rustling sounds from two directions amidst the dense thickets of *Lantana, Chromolaena, Acacia* and *Dendrocalamus* bamboo. The elephants were moving up the slope feeding on the bamboo and the acacia and we decided to follow one of the groups. For nearly three hours we did not sight even a single elephant for such was the nature of the undergrowth. We crawled through a maze of tunnels of thorny *Lantana* and *Acacia*, tearing our clothes and cutting our skin, and trying to be careful not to bump into the elephants. Finally, we emerged into a nullah where we counted five elephants, all cows of varying sizes.

10.55 a.m.: Two cows are feeding on *Acacia pennata* along the edge of the nullah. Twenty minutes later only one elephant is visible. The rest have again retreated into the bushes. From the sounds they are making they seem to be tearing more of the acacias.

12.00 p.m.: The group is standing under the shade of a tree. I still cannot see them clearly.

12.15 p.m. They have gone to sleep. I can distinctly hear a regular breathing sound. One of them also begins to snore loudly. It is a very human-like snore, only much louder! The snoring is intermittent. One elephant is also flapping its ears.

12.35 p.m.: A muntjac suddenly barks close by. It continues to call loudly for the next nine minutes, at the rate of about fifteen calls a minute. The elephants do not seem to be disturbed. Since they have been raiding the fields for the past few nights they may be very sleepy during the day.

1.17 p.m.: Some of the elephants wake up and move out to begin feeding. The snoring elephant continues to sleep on. One of the cows pulls down a gooseberry tree and begins to tear off its branches.

2.45 p.m.: All the elephants are feeding amidst the bushes. Trumpeting is heard from the other group which is lower down the hill. I begin the trek back to Hasanur.

At 5.00 p.m. the same evening from my bungalow I saw the elephants making their way down the hill side. They were clearly heading towards the Binakanhalli farm. I rushed back to the farm along with Setty.

5.25 p.m.: The first elephant, an adult cow, enters the field. Soon the rest follow. This is the sub-group of twelve animals which was feeding lower down the hill. There are three adult females, four sub-adult females, three sub-adult males and juveniles.

5.35 p.m.: The entire group is standing in a row along the forest boundary just inside the field and feeding on the *ragi* stalks.

Setty and I took up position beside a bush about 100 metres away from the elephants. The sun had already disappeared behind the western hills and the light was fast fading, but there seemed to be sufficient light to attempt a few pictures for the record. I sent Setty to fetch my camera from the jeep parked some distance away.

> Noticing Setty, the two adult cows charged in unison. As we ran back, they came half way across the field before stopping. By now the farm labourers had gathered and began to shout. As we all approached the elephants slowly, one cow retreated while the other stood swinging its trunk and kicking up soil for some time before it turned and went towards the rest. Soon, however, the two cows staged a second charge, this time less determined, with one cow taking the lead. The farm supervisor now appeared on the scene with a gun and when he fired into the air the entire herd quickly retreated into the forest.

This group came into the farm on two subsequent nights before moving away southwest towards the grasslands near Talamalai. I never established the identity of this family group but I presume it must has been part of Meenakshi's clan and that they were heading to join the other herds.

Such aggressive behaviour from a raiding herd is rather exceptional. Unlike some veteran bulls, the family herds could usually be chased away relatively easily. Few herds ventured more than a kilometre from the forest boundary into cultivation, while adult bulls had no qualms about traversing ten kilometres across a cultivated tract, passing between buildings, crashing through a cattle shed, or feeding calmly on jack fruit in the yard of a house while its inhabitants shouted helplessly from inside.

One aspect of my study was to determine how much crops elephants consumed and how important these were to them. The first aspect proved to be relatively easy. When elephants feed on paddy or *ragi*, they uproot a clump of plants, invariably chew off only the upper portion bearing the flowers or grain while discarding the basal part of the stems. I measured the area of the field damaged, estimated the densities of plants in the damaged and undamaged portions of the field, and took the weights of samples of whole plants and discarded basal parts. From these I calculated the quantity of forage actually consumed in a field. By tracing the route of the elephants over all the fields damaged during a night's raid, I was able to arrive at the total quantity of crops consumed during a raid by the bull group or herd and by an average individual. I was able to do this for numerous instances of raiding by both bulls and herds.

An adult bull ate on average between 120 and 175 kilograms fresh weight (a quarter this dry weight) of cereal or millet crop, depending on the village involved, while an elephant from a family herd consumed 44 to 76 kilograms. They could certainly consume much larger quantities: a bull can consume up to 300 kilograms and an average herd member up to 200 kilograms. These higher quantities are considerable when we realize that they represent an elephant's forage requirement for a twenty-four hour period. Thus, after eating an estimated 250 kilograms of *ragi* plants during the night of November 1981, Vinay would have had little feeding to do the next day. He probably retired into a shady nullah and got his quota of sleep for the day before heading back towards the same field that evening.

When the quantity of crops consumed is represented as a proportion of the total diet of elephants, it is shown to be less important, at least in the case of family herds. While an average adult bull obtains 9.3 per cent of the quantity of food required in a year from crops, an average individual in a herd gets only 1.7 per cent of its requirements from cultivation. Crops were most important during the *ragi* season between October and December, when they constituted 22 to 30 per cent of the diet for bulls and 4 to 5 per cent for family herds. The cultivated crops are more nutritious in certain respects than wild plants and, hence, contribute more in terms of quality than what these figures indicate.

Why do elephants raid crops? Very often people talk about habitat degradation and insufficient forage in the wild as reasons for elephants entering cultivation. This is only a small part of the story. Elephants have probably been raiding crops ever since *Homo sapiens* took to agriculture in elephant territory. The *Gajasastra* (6th to 5th century BC) alludes to the ravages wrought by elephants in the kingdom of Anga. Kautilya's *Arthasastra* (300 BC to AD 300) talks of the need to protect cultivated crops from depredation by wild elephants. The ancient Tamil Sangam literature describes the male elephant leaving its herd and raiding crops in the hill forests. The reasons for raiding crops go much deeper than proximate explanations such as elephant migration or habitat degradation.

There is certainly a connection between raiding and the

natural movement of elephant herds. Some raiding seems to occur when elephants have to go to a pond inside cultivation or a reservoir in its vicinity. When large herds traversed the maize fields of Kolipalya they headed towards the Suvarnavati Reservoir. Bulls often drank from village ponds when they came to feed on crops. The situation is all the more aggravated when river valleys are taken over by agriculture, because traditional water sources and feeding grounds are then denied to elephants. It is, of course, difficult to say in these cases whether it was the water or the crops that was the primary motivation for the elephants entering a settlement.

Similarly, the fragmentation of the elephant's habitat increases the interface of the forest–cultivation boundary and increases the chances of elephants encountering agriculture in the course of their movements. A classic example of this is seen in the Bannerghatta forests, south of Bangalore city. This long, narrow and convoluted stretch of dry jungle allows little room for elephants to move without touching cultivated land. On 28 January 1985 the students of a college in the suburbs of Bangalore woke up to find nine elephants outside their hostel building. The herd had left the Bannerghatta forests and wandered through fifteen kilometres of settlements before settling down in a mango grove inside the college campus. The forest department had to be called in to chase the herd back to where it belonged.

Ultimately, however, crop raiding can be thought of as a strategy to obtain the most nutritious forage. Evolution has shaped feeding behaviour such that an animal seeks out the best available food resources. Analysing the nutrient content of various foods, I found that cultivated paddy and *ragi* plants had higher amounts of protein, sodium and calcium than do the wild grasses consumed by elephants during this season. Here again, the proximate cause for elephants prefering crops is their greater palatability. Succulent *ragi* plants or sugar cane are surely tastier than the coarse, siliceous grasses found in the forest. Like humans, elephants prefer cake to plain bread!

We can thus expect that elephants will be 'instinctively' guided in their choice of food to plants that will contribute to better health and their survival. Thus elephants would raid crops to a certain extent even if abundant forage is available in the wild, simply because crops are tastier and more nutritious. A bull elephant, having exclusive possession of a large forest territory, will probably head straight towards the *ragi* field once the flowering plants send out their tempting odour.

Why are male elephants more keen crop raiders than female elephants? Putting together all my data, I calculated that a bull entered cultivation forty-nine nights in the year compared to eight nights on the average by a family herd member. Elephants have to face hostile farmers who may throw fire-balls at them, electrocute them with live wire fences, or shoot them dead. Male elephants are thus taking more risks in order to feed on crops by raiding more often.

The answer to this difference between the sexes in taking risks for obtaining nutritious forage may lie in the organization of elephant society. Elephants are polygynous animals. Bulls compete intensely for mating with cows. Some bulls sire a large number of offspring while other bulls may fail to reproduce at all. On the other hand, the cow elephants contribute more or less equal number of calves to the next generation.

Such a difference in the breeding pattern between the sexes—a high variance in reproductive success among males and a low variance among females—is the impetus for the males to take more risks. Risk-taking behaviour would evolve in males if such behaviour promoted their reproductive success. If a bull has to dominate in the mating game it has to be bigger and healthier than other bulls. Dimorphism between the sexes, in this case the males being larger in body size than females, has arisen from this need in the course of evolution. If a male has to grow big and healthy it needs good nutrition, whatever be the source. This would also help the male come into musth, when it gains dominance over other bulls and better access to females for mating. As I explained earlier, mahouts have traditionally controlled musth bulls by reducing their food intake so that they might come out of musth more quickly. All this means that good nutrition is so important to a male elephant that it is willing to take a lot of risks to obtain it.

This is not to suggest that good nutrition is not important for female elephants; they too have their needs while pregnant or suckling their young. It is just a matter of who is willing to take more risks for getting it. In the case of females they not only have to evaluate the risks to themselves but also to others in their herd who are closely related.

In many mammals, the nutrition and growth during the

juvenile years determine the ultimate body size attained. Differences between males and females in feeding behaviour are seen even at an early age. In elephant seals (no relatives of elephants!), the male pups indulge in sneak suckling from cows which are not their mothers to a far greater extent than do the female pups. By doing this they run the risk of being bitten by the strange cows. The nutritional benefits they obtain from the extra milk apparently outweighs the costs of being injured.

Until a male elephant disperses from its family, which is usually at between ten and fifteen years of age, it has to be largely content with following the matriarch. It is only after leaving the family that it steps up its forays into cultivation. Male elephants continue to increase in height until thirty years of age and put on weight even beyond this age. They could thus translate good nutrition into a larger body size even after attaining puberty.

I am not suggesting here that crop raiding bull elephants actually enjoy better reproductive success than those which do not raid crops, but merely that the 'rogue' bull is a product of Darwinian evolution. Elephants face a real risk of being injured or killed while raiding. Pre-historic cultivators possibly speared them or shot them with poisoned arrows, while modern man guns them down. Human selection against crop raiders would have varied through history. Changes in technology, religious beliefs, and political laws would have determined the intensity of selection against raiders. The worship of Ganesha, the elephant god, is a strong deterrent against killing elephants in India. Farmers in some part of India even today believe that the footprint of an elephant in their field would increase crop yields, though such religious taboos are fast fading in the country. Modern conservation laws, however, make it difficult for farmers to kill elephants even if they damage crops.

In any case, agriculture originated too recently to have shaped elephant feeding behaviour through selection. Vinay and the Akkurjorai Bull, the most notorious raiders in my study area, were certainly large and healthy bulls. But it would be premature to speculate at this stage on their reproductive success. Perhaps in the near future the technique of DNA finger-printing will be used to determine the paternity of elephant calves and thereby to measure the breeding success of bulls in a wild population.

When Charles Darwin wrote in *The Descent of Man* (1871) that 'No animal in the world is so dangerous as an elephant in *musth*', he was stating only part of the story. When captive bulls come into musth they are uncontrollable, but this is a very unnatural situation. A musth bull wants to actively seek out a cow in heat, which he is prevented from doing when captured. A wild bull in musth usually does not bother about people; he is too occupied with his lady love or with warding off rival suitors. The image of the 'rogue' bull persists, nevertheless, because bulls are responsible for an overwhelming majority of the deaths caused by elephants. Every elephant region has its stories about the lone bull which lurks near a forest path, terrorizing passers-by, poking its tusks through vehicles and pulling down huts during the night.

During my surveys in southern India I questioned villagers about instances of manslaughter by elephants and obtained over 150 records of such events; of these, eighteen cases occurred in my study area during the two and a half years of my stay there. Most of the people killed were men. This is simply because men come into contact with elephants more frequently. Nearly half the cases occurred within cultivation; in almost all these cases a bull elephant was the culprit. Though dead men tell no tales, there were, however, eyewitnesses to some of these incidents. From these reports I found that male elephants were responsible for over 80 per cent of the killings, females for 10 per cent, and a herd member, not necessarily a female, for the rest.

I have already mentioned earlier how on my first day at Hasanur I came across a man killed by an elephant on the road to Dimbam. The village register of Hasanur mentions another case of manslaughter near the same spot.

> On 14th February 1939 four persons hailing from Salem were travelling from Chamarajanagar on the road to Satyamangalam. Vaithia Pandaram and Chinnapayyan Gounder were walking alongside their bullock cart. Two miles to the south of Hasanur a tusker came on to the road from the east and began chasing them. Vaithia Pandaram was killed by the elephant.
>
> The next day the Inspector of Police retrieved the body and recorded that 'there were gored injuries on the lower abdomen and the left side of the chest. Many ribs were broken . . . ' and added with unintentional humour '... and there were no other injuries on the body.'

Vinay almost claimed a victim at the same place more than four decades later. Assistant Conservator Natarajan, forester Palaniswamy and two guards had been walking along the road to Dimbam when Vinay suddenly burst out of the bushes towards them. They were literally within trunk distance, but fortunately Vinay could not immediately come on to the road because of an embankment along the roadside. Those precious seconds gained before Vinay could reach the road enabled the men to escape.

I was having lunch at Hasanur when the badly shaken party returned to tell me of what had happened. I knew that there was a large pool of water on the rocks in the valley near the spot where this group had encountered Vinay. There was a good chance that Vinay would have gone down to the pool to cool off. We all set off in my jeep along the Dimbam road. When we reached the spot we found that, as I had expected, Vinay was lying contendedly in the pool of water, a luxury for him at this time of the year. Cautioning the others to stay behind I slowly went down the steep path that led down to the pool. I did not want to attract his attention, or to slip down the path into the pool beside him! I was half way down when Vinay abruptly got up and rushed out of the pool, obviously having sensed me, pausing briefly to shake the water off his body. I got just one picture of him facing me before he turned and briskly climbed up the slope of the opposite bank.

Vinay was very suspicious of people and it was very difficult to photograph him. Unlike the gentle Divergent Tusks he would never loaf around or feed near the road during the day. Perhaps his aggressive nature made him distrustful of people.

The Akkurjorai Bull had a long history of killings in the study area. I have earlier mentioned two incidents which took place in 1981 after I began my study. He was not finished yet. The next year, in April, Chikkanashetty and his brother-in-law from Neydalapuram were returning from the forest in the evening. The Akkurjorai Bull had already moved in from the jungle and was standing under a tamarind tree in a fallow field, waiting for all trace of twilight to fade before tackling the coconut grove nearby. Unfortunately for Chikkanashetty, who was walking in front, the path he came on passed directly beneath the tamarind tree. He was caught and flung to the ground. Almost exactly a year later the Akkurjorai Bull stuck again. Chellappan was walking through the fields near Talavadi when he was gored by the same bull.

At the foothills of the Nilgiris, wedged between the town of Gudalur and the steep hills, is the Ouchterloney Valley or O'Valley for short. It has a remnant moist forest which was more widespread before encroachment, cultivation, tea plantations, and the imperatives of resettling repatriates from Sri Lanka ate up a large chunk. Bull elephants living in this tract often come into conflict with the estates and gardens near Gudalur town. One bull killed five people in separate incidents during December 1982 and January 1983.

There are exceptions to the general rule that the male elephant is the more dangerous sex, but these exceptions seem to occur under unusual circumstances. Beginning early 1984, some three or four herds of elephants, totalling thirty to forty individuals, dispersed northeast from the dry forests of Hosur, which are contiguous with the extensive habitat in the Eastern Ghats, through cultivation and small forest patches to the jungles along the borders of the states of Tamil Nadu and Andhra Pradesh. In the latter state, wild elephants had been eliminated by the nineteenth century. Naturally the villagers here were blissfully unaware of the dangers of approaching wild elephants closely. They were extremely curious about these living manifestations of the lord Ganesha, in spite of them damaging their crops. One man went up to offer bananas to the elephants, another fell at their feet in worship. As the human casualties mounted, there were increasing demands for the government to control the elephants. During September to October 1987 one herd of nine elephants was driven back 92 kilometres to the Hosur forest. But the remaining elephants still continue to be in conflict with the farmers. Over thirty people have been killed. Most of these killings can be attributed to female elephants (there were, in any case, no adult bulls among these elephants). In recent years the conflict has reduced as the elephants have settled down in their new home and the people have learnt from their mistakes.

In other parts of India, too, the number of human deaths may be considerable. D. K. Lahiri-Choudhury, who has studied the elephants of northeastern India, informs me that in the state of West Bengal over 250 people were killed by elephants during 1987–90. Considering that this state has less than 100 wild elephants, the human casualty rate is very high. In fact, crisis has recently arisen there. Herds straying from the state of

Experiencing musth for the first time, a confused Biligiri splashes in the pond at Kyatedevaragudi.

With musth streaming down his cheeks, Tippu disappears into the bush on the road to Dimbam.

Setty, my Irula tracker, in the jungle near Hasanur.

The Moyar river which divides the Talamalai plateau from the Nilgiri hills in the background.

Makhna, who is beginning to come into musth, demonstrates in his usual manner.

Vinay rushes out of the pool, pausing briefly to shake the water off his body.

Dodda Sampige, the oldest bull in my study area.

Meenakshi, the matriarch with a calm yet commanding personality.

Meenakshi's son Appu when he was about three years old.

Meenakshi and her joint family having a bath and a drink at the Karapallam pond.

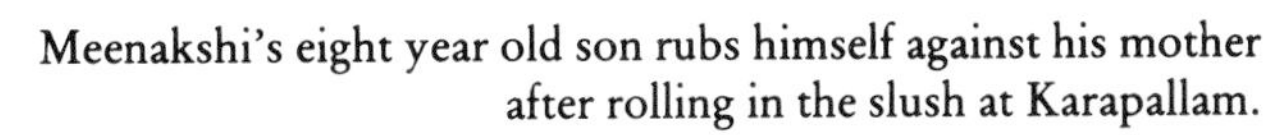

Meenakshi's eight year old son rubs himself against his mother after rolling in the slush at Karapallam.

A cow blows water into its mouth at the Karapallam pond.

'he Karapallam Bull with his left tusk
acked with clay from goring the soil.

Cooling off with a fountain of water.

Kali, the cow with a lump on her forehead, and her three year old son near the Karapallam pond.

A family in search of food and water on the dry Araikadavu bed.

A young bull examines a cow for oestrus.

He then places his trunk and tusks on her back as she turns around.

He then mounts and copulates.

Bihar have been entering the southwestern region of West Bengal and have become a menace to both crops and human life.

There is no doubt that many of the deaths occurring inside the forest are due to the carelessness of people. A cattle grazier taking an afternoon nap beneath a tree beside a stream is simply asking for trouble. Some of the graziers or wood-gatherers who venture into the jungle are old men or women who can hardly walk, much less run, if they encounter an aggressive elephant. Farmers who keep watch over their fields at night from a flimsy hut on the ground, rather than from the safety of a tree, are also highly vulnerable. These are the stark realities of life for the poor living within elephant jungle. I have also known of mentally unsound people, a drunken man who was trying to hold a wild bull by its tusks (he survived!), a deaf man who did not hear an elephant coming from behind, and even an amateur photographer who clapped his hands to attract an elephant's attention, being attacked. Nevertheless, many others who have fallen victim to elephants have been normal people with their faculties intact. While walking through a dense jungle along a path which was created by elephants in the first place, there is neither time nor place for escape if one suddenly chances upon an elephant. Labourers returning from the forest at dusk, women out gathering wood, pilgrims visiting one of the numerous shrines situated inside the forest, are all potential targets for an aggressive elephant.

Since elephants are not carnivores why are they aggressive towards people? The potential for aggression is coded in the genetic make-up of an individual or a species. This operates through the various hormones which are produced and through their action on the nervous system. It is well known that a rise in the levels of testosterone, the male sex hormone, results in increased aggressiveness in many animals. With elephants it is certainly true that the male shows more aggression towards other males. It is also more likely to kill people as a result of its aggression. Male elephants in musth are also more aggressive during this period, when their production of testosterone increases dramatically.

Environmental influences and learning are also important in the development of behaviour, including aggression. The family setting in elephants provides ample scope for the young ones to observe and learn from the aggressive displays of older animals towards predators and people. Aggressive behaviour may also be gradually learnt by a young elephant in the course of suffering pain during competition or play-fighting. Elephants are frequently injured by farmers or by hunters. Those like Vinay, which have broken a tusk, may suffer from a persistent 'toothache'. Such elephants are likely to turn rogue.

The intensity of an elephant's aggressive response can be expected to have been moulded by its past interactions with people. It can also be expected that elephants would be more aggressive towards people in regions where they have been harassed by hunters and encroachers. However, different members of the same herd, which are under similar pressures from people, have very different temperaments from each other, from the very timid to the very aggressive. What makes one elephant tolerate the presence of humans while another, or the same elephant on a different occasion, will attack even without provocation? We can only speculate that an elephant's inbuilt traits (genotypes) shaped by its various episodes of learning during interactions with people (environmental influences) give rise to a wide range of behavioural responses (phenotypes).

From an evolutionary perspective, aggressive behaviour towards other species could have arisen in elephants as a defence against predators. Young mastodonts and mammoths, extinct relatives of elephants, certainly fell prey to carnivores such as the sabre-tooth tiger, the sabre-tooth cat, and the scimitar-tooth cat.

Although adult elephants are too large to be tackled by any predator, the juveniles face the risk of being attacked by lions or tigers. The defensive posture taken up by an elephant herd, with the adult cows bunching together to protect the juveniles amidst them, and the impressive threat displays will have evolved as a response to this threat.

People have also been predators on the extinct relatives of the elephant since the Pleistocene period, perhaps as early as 80,000 years before present. At a site in what was Czechoslovakia the remains of prehistoric humans and the bones of over 900 mammoths have been unearthed, a testimony to the former's hunting abilities. Cave paintings in Europe depict scenes of the hunting of mammoths 20,000 years ago. Modern man has continued to be a predator of elephants, capturing Asian elephants in large numbers for domestication and slaughtering African elephants for ivory and meat. It is thus easy to see that the elephant–human relationship has always been one of prey–predator in the eyes of the elephant. Aggression towards people by elephants can, therefore, be thought of simply as anti-predatory behaviour.

In addition, the more frequent interaction of male elephants with people in the course of raiding crops may itself lead to manslaughter. Most of the deliberate killings within human settlements were by bulls—and these were close to half the total number of cases recorded. Bulls which are aggressive towards people in the crop fields may also show similar behaviour inside the jungle.

The dawn of the new year, 1982, saw the farmers busy harvesting their *ragi* crop. In spite of the considerable quantities consumed by elephants they still on the whole reaped a good harvest. The rain god had been relatively kind that year. All the villages when totalled together lost not more than 10 per cent of their crop to elephants. Some farmers, of course, lost their entire crop while others escaped unscathed.

By January some of the clans had regrouped for their migration to their dry season range. During the first week, when I was briefly away in Bangalore, Setty saw one herd of eighty elephants near Kolipalya, the largest grouping seen during my study. This was probably a good part of Champaca's clan, a few days before they left the plains for the hill forests to the north. Meenakshi's clan was further south, not far from Talamalai.

Many of the bulls also wandered away from the villages at this time, but some, like Vinay, still waited out of sheer habit, to test the patience of the farmers. They came in search of the harvested plants stacked out in the open in large bundles (called *kutharis*) to dry.

One night I was camping in the forest bungalow at Talamalai when I was woken up by a group of excited farmers. A large bull had approached the *kutharis* and was merrily feeding from them, not in the least bothered by their shouting. They seemed confident that their '*yanai* doctor', as they referred to me, could deal with this problem. I was not so sure about this but nevertheless loaded them all into my jeep and drove towards the elephant over the harvested fields. A big black shape loomed out of the dark. The jeep's six blazing head lights and the blaring horn fortunately proved to be too much for the bull; he turned and fled back to the jungle with a yelping sound, rather like that of a dog that has been bitten in a fight. I returned to the bungalow, my reputation intact, to catch up with my sleep.

Another medium-sized bull with a beautiful pair of symmetrical tusks used to frequent the small stream behind my Hasanur bungalow in the evenings before crossing over into the fields. Once the harvest was over he turned his attention to the ripening fruits of tamarind trees close to the bungalow. After he had finished all those within easy reach, much to the consternation of Subramaniam, my cook-cum-watchman, he would rear up on his hind legs and bring down the higher branches laden with pods. One night in January, sitting on the steps in front of my room I watched him shaking the tamarind tree a few metres away, picking up the fallen pods one at a time with his trunk. Noticing me, he slowly moved away down the road in front of the bungalow and into the bushes. In those days I did not have a flash-gun by which to take his picture.

An hour later, I had hardly retired to bed when I heard a loud trumpet, followed by the sound of running feet. Before I could reach the door there was urgent pounding, with shouts of '*Saar, saar*!' As I opened the door Subramaniam and one of the forest guards rushed in gasping '*Yanai, yanai*!' Along with another forester, they had been coming up the road with the dinner of the District Forest Officer, M. Ramachandran, who was camping in the adjacent bungalow. Their lights suddenly picked up a tusker standing close to them in the bushes. The elephant let out a loud trumpet and the forester ran off down the road, while the other two men ran up to my room, the nearest place of safety. The bull, normally of a calm disposition, had probably been as startled upon seeing their light flash as they had been frightened of him. He had retreated into the bushes without any intention of charging. The miracle was that the Forest Officer's dinner was still intact in Subramaniam's hands, so he did not have to retire hungry that night.

It may have been a different story had the elephant been Vinay. A few years earlier Vinay had chased a forester who jumped on to the verandah of the larger bungalow, whereupon he had pushed down one of its pillars. The roof of the verandah collapsed upon Vinay, who then retreated back into the jungle. The broken pillar still lies in front of the bungalow, a reminder to visitors that they had better think twice before taking a stroll in the nearby jungle.

CHAPTER 6

Growth and Family Life

> Tender, copper-coloured, with soft down on his fore-limbs, drowsy, marked by a blotchy trunk, having limbs undeveloped in form, seeking the breast, in the first year he has the name of *bala*.
>
> With toenails getting thicker, with the tongue, lip and the rest (the seven 'red parts') very red, drinking little milk, somewhat inclined to eat creepers and grass, reddish between the foreparts; he capers constantly for no special reason, is generally frolicsome, very fond of sugar, with down-turned eyes, causing delight to the sight, in the second year he is a *puccuka*. . . .
>
> Up to the twelfth year his age makes him worthless; before the twenty-fourth year he is of middling value; up to the sixtieth year this noble elephant is called the best in respect to age.
>
> Nilakantha in *Matanga-Lila*

The morning of 31 March 1982, Setty and I were on our way to Karapallam. During the previous two months we had been busy here once again watching elephants, after several months of trudging through trampled *ragi* fields and listening to the woes of the farmers. It was over a year since I had arrived at Hasanur. Now, looking back at that period in retrospect, I feel that I learnt as much about elephants during that year as I have since then. It had been an eventful year and I still have vivid memories of all my early encounters with these captivating creatures. I can recall exactly where I saw the elephants, how many there were, and what they were doing.

The road from Hasanur to Karapallam goes over a small hill about two kilometres before the inter-state border, from where one gets a sweeping view of the Araikadavu valley to the east and the steep granite hills beyond. That morning in March was as dry as it would get, the brown landscape enlivened only by the orange splash of *Butea monosperma* flowers and the evergreen trees lining the Araikadavu. All along the road were the signs of elephant-broken branches of *Acacia pennata, Capparis sepiaria* and *Zizyphus xylopyrus*.

A herd of elephants was moving up from the valley towards the road. I could only see their backs and hear the crackling of the defoliated bush as the huge beasts otherwise moved silently to cross the road. I have often marvelled at how silent a group of elephants can be as they move through the jungle. They seem to glide on their padded soles without any of the tramping that one usually associates them with. The only way to detect elephants in the jungle is to listen carefully for the flapping of their ears, the rumbling of the stomach, or the crackling of dry bush.

I recognized the old cow, who was not particularly tall but who had ears completely folded over at the top, as Tara. She was leading her family of eight elephants, including a young calf of her own, as they briefly paused to pull at a trunkful or two of the dry plants before moving over to the west of the road.

Accompanying the herd was the tuskless bull, about twenty years old, whom I called Makhna. I had seen him with the family three days earlier and he was obviously joining them. His cheeks were lightly stained as he was coming into musth. As usual with Makhna, he kicked up a fuss, swinging his head threateningly and scraping the ground with his foot before moving off with the rest of the group into the jungle. The herd began feeding and moving slowly towards me parallel to the road. The subsequent events unfolded rather quickly.

At 8.35 a.m. a six year old cow crossed the road from the west to the east, and without noticing me came very close to my jeep which was parked near a small bridge. When it was almost within touching distance it panicked and, emitting a distress call, disappeared into the bushes. Immediately there were deep rumbles from the rest of the herd. Setty and I jumped out of the jeep and went under the bridge to watch. Five elephants including Tara, her young calf and Makhna came rushing out to see what was wrong. They stood at the edge of the jungle, bunched together with their trunks up, the tips swaying from side to side like the raised hoods of cobras, trying to locate where the danger came from. Upon seeing them, the young cow hurried across the road with a squeal of relief and the herd went back into the jungle.

Makhna, however, stayed back, pacing up and down the

road angrily. Finally he began dusting himself with red soil from the roadside. Whenever he heard the sound of an approaching vehicle he would wait until it came up to a curve in the road before quietly hiding in the dense acacia bush. After the vehicle passed by he would resume the dusting. Three vehicles passed by him without noticing him. The third time he made a sound that I can only describe, at the risk of sounding too anthropomorphic, as one of sheer exasperation. He was obviously a very shrewd creature.

At 9.05 a.m. the herd led by Tara came back to the road and they all went down to the Araikadavu.

This is a classic example of how an elephant herd would react when one of its young members was threatened. Even the otherwise peaceable Tara had a streak of aggression in her, and rushed to the aid of one of her family members. This, after all, is what elephant society is all about.

In 1982, the same herds I had seen in the Araikadavu valley during the previous dry season moved there somewhat earlier. From the first week of February I began regularly seeing Meenakshi, Tara, High Head and the Mriga sisters, and their herds, in the valley. They stayed there until the end of April or early May as usual. This was another productive period of elephant watching for me. The Karapallam pond did not attract elephants as frequently as it had the previous year because it did not hold as much water, but this did not really matter. The herds were there in the Araikadavu valley in full strength, busy moving between the practically dry stream bed and their favourite feeding grounds, stripping the bark and breaking the stems of *Acacia suma* at Karapallam.

I too was kept busy observing and photographing them. My file of identified elephants grew and I was pleased to see that the measurements I had made of adult elephants the previous year matched closely with those made this year, while the juveniles grew as expected.

Some bulls which I had not seen earlier began appearing in the Araikadavu valley. Hardly a kilometre from Hasanur I photographed a large bull with a blunt tail and a pair of short tusks that looked very lethal indeed. When I calculated his height I was shocked at the figure; he stood 322 centimetres at his shoulder. None of the other bulls either here or at Mudumalai came anywhere near this size. At first I could not believe my estimation, so I went back to the location where I had photographed him and once again measured the distance (I knew precisely which path he had come along and from where I had taken the shot). The figure was correct. I turned to the prints again and recalculated the height. The height was indeed accurate. Could this have been the much-dreaded Akkurjorai Bull? I had never actually seen him before. If it was him, what was he doing near Hasanur? To these questions I simply did not know the answers.

Two other bulls, Cradle Tusks (who also had a blunt tail) and a younger bull with beautiful incurved tusks which I called Cross Tusks, also came to the valley. One busy morning Setty and I had photographed a number of herds, as well as Cross Tusks who was trailing the Mriga family. He was in musth and in some hesitation came up to us before disappearing into the jungle behind the others. After the coast was clear of elephants, we got out of the jeep, took out a tape and began our distance measurements. We were engaged in this for a few minutes when, glancing back, I noticed Cross Tusks approaching the jeep from the opposite direction! We retreated in a hurry not wishing to mess around with a musth bull, however harmless he might seem. Cross Tusks slowly walked up to the jeep, stopped in front of it, and ran the tip of his trunk over the windscreen. Then he moved over to its open side and put his trunk into the jeep. My camera (and his undeveloped picture) was lying on the seat and I wondered if he would become the first elephant to trigger the shutter and take his own picture! Not satisfied, he went to the back of the jeep and again stuck his trunk inside, no doubt absorbing the human smell that still lingered there. His curiosity finally satisfied for the moment, he walked over to an *Acacia suma*, pulled down a branch and headed to the shade of the Araikadavu.

I was to eventually meet Cross Tusks at Kyatedevaragudi that November but, apart from him, I did not see any of the other elephants I had identified in the Araikadavu valley whenever I went north to the Biligirirangans. I could have certainly missed some of the bulls, but all the herds I saw here seemed to be different ones. These ranged over the moist deciduous forests of the BRT Sanctuary during the dry months and the early wet season. Of these, Champaca, with a large tear in her left ear, was the most easily recognizable. I could recognize a few other groups by identifying a large cow or a sub-adult bull with peculiar tusks, but I never got around to naming them as my sightings of them were not too frequent.

By now a picture of how elephant society was organized was beginning to crystallize. I was, of course, familiar with the descriptions of African elephant society based on more detailed research and I found that Asian elephants were organized in similar fashion. There were differences in specific detail, such as group size and so on, but basically the structure was the same.

Elephants live in families led by the oldest female. The shikar literature is replete with references to the alleged 'master bull' of an elephant herd. If the term master bull implies that a bull is the unquestioned leader of a herd then it is no doubt a figment of the male chauvinist's imagination; the human male, and that too a macho hunter, cannot perhaps think otherwise. This is surprising because, as early as 1878, G. P. Sanderson had observed that 'an elephant family is invariably led by a female, never a male'. The truth is that among elephants it is the lady who gives the marching orders. No bull, however formidable he may be, can usurp her authority.

A family typically has one adult female and anywhere from one to five immature children. The term family has also been applied to larger groups. In a group there may be three generations, where the matriarch has mature daughters who have their own offspring. A group may also consist of two or more adult females, presumably sisters, and their children.

Ian Douglas-Hamilton felt that elephant groups with more than one mature female could be termed 'extended family units', but he retained the term 'family' in view of their cohesion. In Lake Manyara he found that the extended families were remarkably stable. Whenever he saw an identified family its composition was practically the same. Later studies in Africa, including those of Cynthia Moss at Amboseli and Rowan Martin at Sengwa in Zimbabwe did not confirm this, however. The splitting and fusion of different families were common even over short time intervals.

In southern India, both in the Biligirirangans and later in Mudumalai, I observed that the only really stable elephant unit was that of one mature female and her immature children. The matriarch Meenakshi sometimes moved with only her immature daughter and two sons. At other times she was seen with her entire family, including her two mature daughters and their children. Presumably her mature daughters also spent time on their own with their children. On yet other occasions Meenakshi and her family, larger or smaller, were seen in the company of another family led by the old cow Tara, who may have been Meenakshi's younger sister. Other well-identified families also behaved similarly. Champaca either moved along with only her three children, or with a larger group consisting of her sisters (presumably) and their offspring.

In view of these observations I propose that the term 'family' should be restricted to a single adult female plus offspring, and the term 'joint family' is used, in the Asian tradition, to describe groups with more than one adult female, even if these are reasonably stable.

Above the family and joint family are other levels of organization. A family or joint family may show strong ties with one or more families in the area. These may well be related families which have separated at some time in the past. Douglas-Hamilton called them 'kin groups'. If relatedness is not necessarily implied it is better to refer to them as 'bond groups', as Cynthia Moss does.

A still higher level of organization seems apparent among elephants during the dry season. While searching for elephants at this time of year I noticed that if I came across a herd in an area there would almost invariably be many other herds nearby. At the same time, there were other areas that were practically free of elephants. In other words, elephants distinctly congregate in certain places. Each of these clusters, made up of several elephant families, seemed to represent a 'clan', as has been described of some African elephant populations.

In my study area each aggregation had between 50 and 200 animals. To the north, the clan I named after Champaca, ranged over the moist deciduous forests of the BRT Sanctuary. Meenakshi's clan, meanwhile, occupied a more central location in the Araikadavu valley. From density estimates in the valley during the dry months I figured out that about 125 elephants made up her clan. A much smaller clan was seen to the west near Talamalai. A large aggregation of some 200 to 300 elephants occurred early during the dry season to the south in the Moyar valley and the jungles around Bennari, though there may have been more than one clan there. Another clan ranged to the east in the vicinity of Gaddesal. There were thus five or more clans which used my study area at least part of the year.

Indeed, the families constituting a clan seemed to be broadly co-ordinated in their seasonal movements. The home ranges of adjoining clans did overlap to some extent, though not necessarily during the same season.

These clans were part of the larger population of over 5000 elephants that range over the Nilgiris and the Eastern Ghats.

Unlike female elephants, who remain with their families even as they grow older, the male elephants leave their families when they are on the threshold of sexual maturity, usually between the age of ten and fifteen years. The process of separation is a gradual one. The young, maturing bull spends more and more time away from its family, either alone or in the company of other young bulls, until it has for all practical purposes separated from its family.

One young male in Tara's family, which I named Dancing Bull (because of his peculiar manner of shuffling sideways whenever he saw me), seemed in the process of breaking away from his family in 1982 when he was an estimated thirteen years old. One of Meenakshi's sons, who was invariably attached to her during 1981 and 1982, was not seen with the family in 1983, when he would have been about eleven years old. Although I have never seen it happen, others have reported that the adolescent bull may even be physically pushed out of the family by its mother.

In terms of the survival of the species, it would be adaptive for the members of one sex to leave the family into which they were born and to go elsewhere to breed. This would lessen the chances of close relatives mating and producing children. Inbreeding can have many undesirable consequences. The offspring have greater chances of inheriting genetic defects, while the population suffers a loss of its genetic diversity.

After leaving their families, the bulls wander on their own or seek the company of other bulls. There is no evidence from any study on elephants to indicate that associations of two or more bulls are anything but temporary. No special bonds seem to be formed by the bulls in an all-male group although, as I had observed earlier, two or more bulls may associate for up to a month at a time for raiding agricultural fields.

Most of the adult bulls I saw in the jungle moved alone. There were very few times I saw two bulls together and just once I saw three young bulls keeping each other company. The large bull groups described of many African elephant populations did not occur. This was partly became there were, relative to the total population, far fewer adult bulls in my study area, ivory poaching having eliminated many of them. Even in Sri Lanka, George McKay had observed bull groups of up to seven animals. The males here, which were mostly tuskless and, therefore, immune to ivory poaching, constituted a much larger proportion of the total population than do those in southern India.

Adult bulls join the family groups temporarily for mating with oestrous cows. Here again there is no evidence that a bull always associates with a particular family group. The associations are opportunistic and depend upon other factors, such as dominance hierarchies among bulls or even the choice of the cows. The adult bulls spend 20 to 25 per cent of their time on average with family groups.

Since male elephants cannot 'recognize' their children, they show no interest in taking care of the young. The opposite is true of families. The members of a family form very intimate bonds. Mother–child ties are the strongest, but close bonds do exist among all the members. Not only the mother but also her sisters and older siblings participate in the care of the young. If an adult cow does not have a young calf of her own she may even allow another's calf to suckle, an act termed as 'allomothering'. Such altruistic behaviour would be in an animal's own self-interest. By helping a close relative, with whom it shares a significant proportion of genes to survive and reproduce, the seemingly altruistic individual also ensures that copies of its own genes are passed on to the next generation. This argument, known as kin selection, was first explained in clear mathematical terms by biologist W.D. Hamilton in 1964.

Elephants have evolved a complex and sensitive system of social interaction. Members of a family or bond group communicate through touch, sound, and scent. It is common to see elephants rubbing their bodies or pressing their foreheads against each other, placing their trunks into each others mouths, intertwining trunks, or just going into a huddle. Such interactions may occur in different contexts—when sub-units of a joint family or bond group meet after a temporary separation, when the animals are in a playful mood, or when they are under some stress. These interactions reinforce the social bonds.

When elephants are under stress they may seek reassurance through touch. A common expression of nervousness is seen when a younger elephant goes up to an adult cow, entwines trunks with her and places the tip of its trunk in her mouth. On 7 April Meenakshi's family seemed unusually nervous as it crossed the road to go down to the Araikadavu river near Karapallam. Along with the family were Meenakshi's sister Tara and her younger children.

> 9.05 a.m.: The elephants cross the road facing me with their trunks lifted. After they reach the other side they are still bunched together. Meenakshi is in front with one of her adult daughters to her right. The old cow, who could be her sister, is behind. Her daughter places the tip of her trunk in Meenakshi's mouth. Meenakshi reciprocates. After a few minutes the herd turns around to go to the Araikadavu. Meenakshi now goes up to her sister and places her trunk in the latter's mouth.

Elephants also use a rich variety of sounds to communicate. A low rumble seems to be used for making contact at short distances, while a full-throated roar may be used for calls over longer distances. The latter can be more commonly heard during late evenings or night when sub-groups of a large family may wish to come together.

A series of short squeaks that can be rendered, '*kook, kook, kook* . . .' seems to indicate a state of conflict plus, perhaps, an alarm signal. I have commonly heard this call when I have surprised an elephant when it has strayed from the rest of the herd or when it has retreated after a half-hearted charge. A more determined charge may be accompanied by a shrill trumpet. An elephant making a threat display may scrape its foot on the ground, at the same time hitting the ground sharply with the tip of its trunk, which produces a booming sound not unlike that of a muffled explosion.

The most intriguing of elephant vocalizations is the throbbing, '*phut, phut, phut* . . .' like the sound of a motorcycle. As I previously mentioned, I really did mistake this sound for a motorcycle when the Karapallam Bull was coming up from behind my back! Recent research indicates that this is merely the audible part of an infrasonic vocalization.

As early as 1972 the Indian naturalist M. Krishnan stated that elephants may communicate at sound frequencies that may not be fully audible to humans. He had variously described these sounds as 'throaty, hardly audible', 'low-pitched but clearly audible from a distance', and 'a throbbing purr'. These are very apt descriptions of the phenomenon. Unfortunately, he did not have the necessary instruments to record and analyse the sound frequencies.

It was left to Katharine Payne and her associates working at the Washington Park Zoo in Portland, Oregon, to record and precisely characterize infrasound. While observing Asian elephants at the zoo one day in May 1984 she sensed a throbbing sound that puzzled her. Later the thought occurred to her that it may have come from the elephants, which could be communicating among themselves with calls that the human ear is incapable of hearing. She had been trained in music and had earlier worked on infrasonic communication in whales. With the aid of recording instruments she confirmed that certain calls of elephants have frequencies that may be as low as 14 hertz. A human with perfect hearing can catch sounds from 20 hertz upwards to about 20,000 hertz; those living amidst the din of modern civilization can probably only hear sounds above 30 or 40 hertz.

The discovery of infrasound among elephants opened up entirely new vistas for investigation. The remarkable co-ordination in movements of related families of a clan could now be explained. A cow in oestrus could silently advertise her condition to bulls in the surrounding area and they would come flocking to court her. By means of experiments, Katharine Payne and her associates actually found this to be true of African elephants in the wild.

This discovery also raised other disturbing questions. Could an elephant herd which was being hunted, either during official culling or by poachers, communicate the message of death to other herds in the region?

Infrasound could perhaps be even more important as a means of communication to the forest-dwelling Asian elephant or the African forest elephant than it is to the savannah-dwelling African bush elephant. Infrasound can travel long distances through forest without attenuation. The same would not be true of scent in dense forest. This could provide a clue as to why the temporal glands of the female Asian elephant do not seem to secrete temporin, that serves as an agent of communication, as do the African savanna elephants'. This communication function of the temporal glands could have either been lost in the course of evolution or not have developed sufficiently in the forest-dwelling Asian species. Actually, as I have mentioned earlier, they do secrete occasionally, and this could be because the populations I have observed in southern India live in habitats that are relatively open, like the savannahs of Africa. The crucial evidence as regards this will come from further studies on the African forest elephant, a sub-species whose biology and ecology is very poorly known.

The elephant family is superbly geared to confront any predator other than man. The older elephants are too big to be seriously challenged by even the largest predator such as the tiger, though the juveniles are certainly vulnerable. Indeed, elephant calves are known to fall prey to tigers. To protect the young ones the older cows in the family make up a formidable defense – offense system. When any danger confronts the young ones the cows can be very aggressive indeed.

A herd may also be agitated by even a small predator such as the dhole or Indian wild dog that hunts in packs of four to eight animals. One morning in August 1982, Tara and four others of her family were standing out on the rocky bed of the Araikadavu, sucking up water from a small pool. The monsoon had failed to live up to its promise during the previous two months and the elephants were in poor condition. Tara's bones stuck out through her skin and her temples bulged prominently. The animals were quite listless as they stood out in the open under a fierce sun. After drinking for a while, Tara put up her trunk in the air, testing the breeze. She became quite agitated, scraping the rocks with her feet and making low growling noises. The herd bunched together, with the calf secure in between the larger cows. I first thought they were agitated after getting my smell, but a few minutes later a pack of five dholes emerged out from the jungle. The elephants wheeled around to face the dogs and even took a few steps forward, but the dogs just turned direction and vanished. They were certainly not after the elephants!

It would be rare to observe a direct confrontation between elephants and a large predator such as the tiger; in most places where tigers can be easily seen in India, such as at the Kanha or Ranthambore national parks, elephants are absent, while in areas where elephants are abundant, the tiger is elusive. One outstanding place for observing both the species together is the

Nagarhole National Park. I was once fortunate enough to witness the interaction of elephant and tiger in the company of wildlife biologist Ullas Karanth, who was studying tigers and leopards at Nagarhole.

Ullas and I were sitting up one of his favourite watch-towers located some 200 metres away from a water-hole. At about 3.00 p.m. a herd of elephants made its appearance at the pond.

'Oh', groaned Ullas, 'your elephants are a positive nuisance! They will keep my tigers away.' However badly I wanted to see a tiger, I could not but disagree.

The elephants did not sojourn for long, but departed after drinking only a few trunkfuls and without even bathing. Soon another herd of nine elephants entered the water and they too came out and huddled together amidst the bamboo clumps, some of them trumpeting half-heartedly. Only a young male, about five years old, still lingered in the water. We thought this to be rather strange as it was a hot day in March at the peak of the dry season. Added to this the spotted deer grazing nearby were sounding an alarm. Ullas was sure that a predator was close by. He knew these signs well.

At 3.25 p.m. Ullas suddenly grabbed his camera, whispering excitedly, 'Tiger!' Sure enough, a large tiger walked up to the edge of the pond in full daylight. The young tusker, who was still drinking, rushed out in alarm upon seeing the tiger. The other elephants did not seem unduly worried as they could now actually see the predator. The tiger did not seem to relish the idea of a sauna with the elephants nearby and quickly disappeared into the jungle.

For once Ullas had not brought along his radio-tracking gear, as he was not expecting one of his collared tigers to be present at that place and, in the excitement, neither of us had noticed whether or not the tiger had a collar or not. In frustration Ullas went back to the camp to pick up his equipment. When he returned half an hour later along with K. M. Chinnappa, the indefatigable Ranger of Nagarhole, he confirmed that the tiger was Das, one of his collared males, for whom we had been searching without success for that entire morning.

At 6.15 p.m., when the pond was free of elephants, Das returned to sink into the warm, relaxing water. Chinnappa could not resist getting down from the tower and sneaking up to the pond through a circuitous path that ran through the jungle. After ten minutes or so Das turned around and began paddling, as if he wanted to come out of the water.

'Watch out, he is going to charge at someone', whispered Ullas. Seconds later, a bull elephant, not more than fifteen years old, charged into the water and Das in sheer frustration turned around to beat a hasty retreat. Upon reaching the edge of the pond all that Chinnappa saw was the elephant drinking victoriously.

The defense–offense behaviour of an elephant herd towards a predator resembles its reaction to a perceived threat from people. The elements of aggression can be described in the classical terms of conflict behaviour: threat display, displacement activity, redirected aggression, attack, and retreat. When an elephant smells or hears humans its immediate reaction is to suspend all activity for a while and to concentrate on locating the source of disturbance. The trunk may be half-raised and the tip moved in an arc to detect the direction of the scent. The ears stop flapping and are held half or fully extended. In most instances, the entire herd retreats in the opposite direction. On occasion, an elephant may approach the source of the stimulus in an aroused state even without seeing a person.

Makhna was a particularly aggressive bull who charged practically every time he saw me or even when he smelt me. On one occasion he had finished a bath and was rubbing himself against a tree on the farther side of the Karapallam pond as I watched him from behind a bush.

The wind was unfortunately carrying my scent towards him. While still rubbing himself, he put up his trunk to get the direction of my scent. Though he had certainly not seen me, without warning he ran around the pond towards me, trumpeting twice only after crossing over. I had retreated to my jeep and was moving away when Makhna burst out into the open. At the same time a bus was coming along the road in the opposite direction. The bigger object caught Makhna's attention and he ran straight towards the bus, which managed to squeeze past him as he came up to the road. Frustrated he began a typical 'displacement activity' of dusting himself in an exaggerated fashion with soil. Finally he went over to a large tamarind tree to resume the activity he was originally engaged in, that of rubbing his hind leg against the trunk, before heading back to the pond.

When an elephant notices a person there may follow a period of conflict. A nervous elephant may place the tip of its trunk inside its mouth or, if a tusker, drape it over its tusks. Elephants may seek reassurance from one another through trunk or body contacts. The elephant may then stage a mild threat display by fully extending its ears so as to appear even larger, swing its trunk rhythmically, shake its head, or sway its whole body. It may scrape the ground in conjunction with displacement activities, such as gathering mud and grass with its trunk and throwing it over its body. Sometimes the displacement activity takes the form of a rapid feeding upon grass.

An agitated elephant may run a few paces first towards and then away from the enemy, making trumpeting sounds, in an attempt to scare the intruder. If this fails it may launch a more serious attack—a mock charge culminating in an impressive display within a short distance from the enemy.

It is not always the largest female elephant or matriarch that puts on this threat display; very often it is a younger cow in the herd. I have generally found the oldest and most experienced cows to be relatively calm, while sub-adult cows were often more agitated and charged.

An elephant may even charge without any warning or threat display. During a determined charge the trunk is coiled inwards and the ears held back close to the neck. For all its seeming clumsiness, an elephant can run pretty fast. I have clocked a charging cow on my jeep's speedometer at about thirty kilometres per hour, although it cannot sustain this speed for long (an olympic sprinter completing 100 metres in ten seconds is doing only 36 kilometres per hour). Considering that an elephant can bulldoze its way through bushes, even an athlete would be hard pressed to outrun a determined elephant over short distances.

At the end of a charge an elephant spreads out its ears and unfurls its trunk, delivering a sledge-hammer-like blow to any object unfortunate enough to be close by. There is considerable variation in vocalization during a charge. A charge may begin and end in complete silence. On occasion there is a warning trumpet before the charge, but more often a shrill trumpet is let out either during or at the end of the charge.

If an elephant decides to retreat after a charge, it turns sharply to one side with its head raised and its tail up, and moves away at a tangent, keeping the enemy in view until it reaches a safe distance. It may then enter into a phase of 'redirected aggression' by thrashing the bushes or breaking a branch from a tree. One cow even broke a dried branch from a fallen tree and flung it in my direction.

Actual physical contact with a vehicle is rare, although this possibility must be always borne in mind. It is wise not to fool around with a three-ton missile which is programmed to seek out and destroy you. A five-ton missile armed with a gleaming pair of weapons, the tusks, can be even more lethal. In the northeastern Indian state of Arunachal Pradesh, an army truck was once pushed off a mountainside by a tusker, resulting in many casualties.

What do you do if you are faced with an angry elephant? If you are on a steep slope, it is better to run down hill as the elephant will be afraid of losing its footing. If there is a ditch sufficiently deep, the best course would be to quickly get across it. As a last resort, an elephant may even stop if you shout loudly. Many of these rules, of course, break down when an elephant is actually chasing you. Then all that one can think of is to merely run away.

I have known of cases where, upon running away from an elephant, the people have fallen down and, expecting to be crushed at any moment, have found to their surprise that the elephant stops near to them and proceeds to cover them with mud and leaves, burying them with due honour! Arun Chandrasekhar, a young wildlife enthusiast from Mysore, had a remarkable escape after being attacked by an elephant in Kerala's Parambikulam Sanctuary in February 1983. After observing an adult bull for several hours from the safety of a watch-tower, he came down in the evening to return to his camp. The bull noticing him charged, caught up with him and lashed out with its trunk. When Arun fell down the bull retreated and stood some distance away. After regaining his senses Arun, who had fractured a collar bone, managed to crawl up to the watch-tower. Fortunately for him, a jeep with some officials soon came by and put him on a bus to Coimbatore. Here he contacted Tilaka and Theodore Baskaran, keen naturalists whose house was a favourite meeting place for all wildlifers in the region, and with their help was admitted to hospital. The very next day the same bull killed a man from the local tribal settlement.

This behaviour of 'burying' the dead has also been described in the African elephant. Elephants are especially curious about the bones of dead elephants and will examine them, pick them up with their trunks and even carry them off some distance. Sometimes they pile up leaves and mud over a carcass. There are also anecdotes about cow elephants standing guard over their dead calves or even carrying the carcass around for several days. We have no reason to disbelieve these accounts. Concern for the dead is not solely a human prerogative. The elephant is a remarkably sensitive creature.

The significance of this behaviour has been a puzzle to elephant observers. I think a perfectly rational explanation is available. I base this upon a brilliant deduction that one of my graduate students, Milind Watve, made with regard to burial behaviour among humans. Milind was actually tackling the question of why tigers indulge in man-eating in certain regions. He approached it by asking the opposite and more relevant question as to why tigers and other predators do not kill many more people than they actually do; after all humans are pretty frail creatures. He reasoned that it was because humans retrieve and bury the dead. It would be a waste of a tiger's energy if it were to spend time in killing people only to find that it was not possible to consume the carcass at leisure without it being taken away by other people. The earliest evidence of burial among humans is seen in the Neanderthals about 70,000 years ago. Once this practice spread it was no more very attractive for predators to kill people. Primitive humans had taken a giant step in freeing themselves from the continual fear of being preyed upon by large beasts. Where man-eating occurs at present times, it is usually in remote places such as the Sundarbans, where dead people cannot be retrieved easily, or in places where 'modern' state laws make it imperative that people wait for the police to arrive before removing a dead person killed by a tiger in the jungle, thereby providing an opportunity for the animal to completely devour the carcass. Man-eating, or avoidance behaviour, in predators could be culturally passed on from one generation to the next.

It is, therefore, reasonable to think that the behaviour of elephants guarding carcasses or attempting to bury them is adaptive. Although a tiger (or a lion) cannot bring down an adult elephant, it can certainly cause panic in a herd sufficient to allow it to grab a young calf that would provide it with meat for several days. However, the fact that adult cow elephants refuse to easily leave a dead calf behind and to move on makes it unattractive for tigers to prey upon elephant calves in the first place. The behaviour of inspecting carcasses or picking up the bones of a dead relative reinforces the strong social bonds that exist among elephants of a family. In the course of evolution such behaviours that improve the chances of survival of a creature would have no doubt been favoured.

The family setting is thus indispensable for the normal growth and development of the young elephants. Within the family, the calves are protected, nourished, nurtured, and taught the rules of living. At birth the calf is a pretty helpless creature and needs all the help it can get from the elders in the family.

To follow the development of various behaviours over time in a particular elephant calf in the wild would be a rather tall order, unless one were to radio-track a family. I had to be content with making casual observations on the behaviour of calves of different ages. Around the time I was beginning intensive work at Hasanur, my colleague Vijayakumaran Nair was concluding a study on the development of behaviour and calf–mother relationships in captive elephants held in the forest camps at Bandipur and Mudumalai. I have drawn upon his observations to fix the precise age at which different behaviours appear.

The first few days after birth the calf walks with an unsteady gait, keeping close to its mother, searching between the mother's legs for its source of milk. When it does find a nipple, it suckles briefly for a minute or two. The herd is considerably slowed down by the calf, for it rests frequently with its mother. Sometimes others stand guard and provide shade. Although unable to pick up anything, the calf inspects its surroundings with curiosity, touching objects with its trunk and placing the tip in its mouth.

After a week, the calf is bold enough to enter a pond along with the rest of the herd. It cannot yet suck in water with its trunk but drinks directly through the mouth. The rubbery, flexible trunk is a source of puzzle to the young calf and one of amusement to the human observer. Wriggling it about, twisting it around in the air, placing the tip in its mouth or even tripping over it, as yet the calf seems unsure of what to do with it.

Mother and calf stick close to each other, in continual contact with each other through their trunks. The mother reassures her baby by running her trunk over its body or by placing it near the calf's mouth. When Vijayakumaran tabulated the various activities of a calf and its mother, he noticed that much of their contact was seemingly irrelevant. In some eighty per cent of cases there was no clear, immediate function that could be attributed to the exchanges between the calf and its mother.

Madhav had a simple explanation for this: the contact was a calf's way of letting its mother know that it was in fine health and a mother's way of confirming that this was indeed true. Human analogies immediately spring to one's mind. A listless child is signalling that it is in need of medical attention. Similarly, an elephant calf that is not fidgety enough may be indicating to its mother that she should slow down her pace of movement, provide it with shade and rest, or pour water over its body.

When the calf is a month old, it can pull up small herbs from the ground but, although it tries to, it is unable to uproot grass clumps. It is able to collect soil with its trunk and throws this under its body. By now it sometimes leaves its mother for short periods. It goes up to other members of the herd, often juveniles, not losing the opportunity to grasp their tails or to climb over them when they are lying down.

At between two and three months of age, the calf is totally at ease in water. It may even rush ahead of the rest of the group into a pond. While bathing the calf immerses itself completely underwater, sucks in water and blows it out of its trunk, and even clambers onto other elephants. The calf still stays close to the other members of the herd, quickly disappearing between their legs at the slightest hint of any danger.

By six months, various other elements of behaviour appear. The calf can manœuvre its trunk sufficiently to blow out water on to its back and sides. It can pluck tender grass or green leaves from shrubs and eat them. Curious about any intruders, such as cattle, it will now go forward a few steps, making a low rumbling sound and spreading its ears, a hint of the full-blown threat display which will develop in later years. However, for the moment, if the opponent retaliates, it will quickly retreat to the safety of the herd.

Before its first birthday it is able to feed to a limited extent on the leaves of most plants. It can pluck the leaves of bamboo, even holding a branch down with its foot, while feeding. Using its front foot to scrap the ground, it collects soil with its trunk for throwing over its body. It may even uproot small clumps of grass, beating them against its legs before eating them. Nevertheless, it depends on its mother for much of its nourishment. When a herd is slowly moving through the jungle foraging, the calf may wander away towards other older juveniles in order to play. Standing 120 centimetres (4 feet) tall, an increase of a foot from its height at birth, the one year old fits exactly beneath the belly of its mother. This is a useful guide to identifying calves below one year of age in the field. Its weight has increased from about 120 kilograms at birth to about 330 kilograms at one year.

During the second year of its life the juvenile is on its way to becoming nutritionally independent. Its ability to pull, twist, and tear plant parts increases substantially. It becomes adept at co-ordinating its trunk and front feet in order to remove soil from grass clumps. It is increasingly playful with its peers and elders in the herd. Butting, pushing, chasing, wrestling with the trunk, pulling another's tail, rushing forward threateningly, the juvenile is an endearing little rogue.

The significance of play in the normal development of behaviour in an animal has been realized only in recent years. Play experiences accelerate the development of the brain and nervous system, which in turn control various behaviours. In higher social mammals the role of learning in the development of behaviour is well known. Cognitive and motor skills, used in interacting with other individuals and with the environment, are sharpened. Play helps in the recognition of kin and the formation of social bonds which will come in useful in later life. In short, play prepares a young animal with the experience and skills that might one day be vital for its very survival when confronted with the unexpected.

A two year old elephant stands about 137 centimetres tall and weighs about 500 kilograms. In a male elephant, the tusks just stick out from below the lip line. The budding tusks can be seen up to six months earlier, but only if the trunk is lifted up, as when feeding. This is again a useful clue when aging elephants in the field.

Analysing the large amount of data I had on elephants born in captivity, I found that there is no difference between male and female calves as to their average height at birth or in their growth rates during the first two years of life. The pattern may, of course, be somewhat different in wild calves. In any case, in captivity the juvenile males clearly begin to outgrow the females from the third year onwards.

The juvenile males begin to explore the world on their own much earlier than do the young females. The young male also seems more intolerant of real or imaginary enemies. My observations one evening in April 1982 illustrate this point.

> Meenakshi and Appu are drinking from a small pool in the Araikadavu. A flock of common mynahs also compete for the moisture. Appu rushes at the mynahs with low growls and chases them away.

The young males also interact among themselves more intensely than do the young females. I do not have quantified data to prove this, but it appears as though the young males organize mini-conferences while the rest of the herd rest or feed. On these occasions, two or three juvenile males get into a huddle a short distance away from the others and begin a session of play.

On the morning of 15 August 1982, I was observing Kali's joint family of nine elephants at the Araikadavu. The elephants were having a difficult time obtaining water, as only stagnant pools now graced the Araikadavu. The animals appeared to be in surprisingly poor shape for the peak of the wet season. This was rather disturbing. If the elephants did not have sufficient energy reserves now, how would they fare in the coming months?

The herd was around a bend in the stream, so Setty and I crept up close to them behind bamboo clumps. The warm smell of elephant mixed with that of wet soil wafted back to us as we watched them try to extract some moisture from the dry stream bed. My notes were as follows.

> There is now no flowing water in the Araikadavu. The elephants are facing away from me. They are digging holes in the sandy bed. One large cow uses both a forward and a backward kick to make the hole. Simultaneously it uses its trunk to suck up the muddy water and with a forward flick spray it over the ground. After doing this a few times it drinks from the hole. Once it also sprays the muddy fluid onto its small calf, which appears to be not more than a couple of weeks old. I wonder whether it will survive the drought.
>
> The drought has not dampened the spirits of the young tuskers. Three young males (three years old) move about twenty

metres away from the rest and begin to wrestle using their trunks. One male goes back but the other two persist in their play. Now the small calf comes up to them and sprawls between them in the sand.

After being unaware of my presence for half an hour, the large cow put up its trunk and got our scent. She made a low rumbling sound and all the other elephants ran up to her. The entire herd immediately turned and silently came up the path from where we were watching them. By now we had retreated. When they reached the spot where we had stood they stopped for a moment and then veered away into the bushes.

Play-fighting intensifies as the males grow older. From simple butting or wrestling using its trunk, the play changes to a test of their relative strengths in pushing with locked tusks or even in the clash of ivory. It is rare, however, for a sub-adult animal to be injured. Through play-fighting an animal learns its strengths and weaknesses relative to other males in the population. Later on in life this may help in determining a male's position in the social hierarchy. When the stage is set for a serious clash, the earlier experiences acquired during play-fighting could possibly help prevent the fight from becoming fatal; a male would realize when to retreat from a more dominant opponent.

As the youngsters grow, they have tremendous scope for learning behaviour from elders in the herd. At birth an elephant calf's brain weighs only 35 per cent of what it will ultimately weigh during adulthood. The lessons learnt during growth are quickly assimilated—which plants can be eaten; when to migrate to the valley lush with bamboo; what to do when threatened by a predator; how to behave when pursued by an ardent bull. The importance of what is learnt by experience and how this interacts with the elephant's genetic make-up can be understood from the variety of behavioural responses seen among elephants. Each elephant is different from every other elephant, not only by virtue of its distinctive genes, but also because it has undergone unique experiences in life.

As more members are added to the family, as the adolescent males depart, as the elephants grow old and die, the family takes on a new look. Once a grandmother dies it is time for her daughters to go their own ways with their own respective families, though they still meet and socialize.

Potentially, a cow elephant can live up to eighty years of age or so. In the wild this age may be only very rarely attained because an elephant would normally die of starvation much earlier on, once all its teeth are worn out.

An elephant develops six sets of molar teeth on either side of their lower and upper jaws during its lifetime. The teeth do not all appear at the same time but do so progressively as the animal ages and the previous teeth are worn out. At any given time there are usually no more than two sets of teeth in use. The teeth appearing later in life are larger, are made up of more lamellae and last longer.

A calf is born with its first molars, called Molars I and II. Molar I is shed quite early on, between the age of one and two years. Molar II lasts until it is five or six years old. By this time Molar III is already in use and lasts until it is eleven or thirteen. Molar IV appears by the time it is nine and is worn out by the time it is twenty-five. Molar V erupts at thirteen years of age and lasts until it is anywhere between forty and fifty years old. Molar VI is the last one to appear, at the age of thirty to thirty-five, and lasts into old age. For the Asian elephant in the wild the last set of teeth seems to last until seventy years of age, or rarely, even longer.

By this age the old elephant seeks the relative security of a river or pond where it can feed on succulent plants and be ensured of adequate water. This tendency for elephants to die near water has no doubt given rise to the myth of elephant graveyards.

Going through my records of over a thousand elephants which were born in captivity or captured in southern India since the late nineteenth century, I found that a cow named Peri holds the record for longevity. She was captured in North Malabar on 11 April 1889, when she was at least 20 years old, and died on 9 December 1948 at the age of 79 years. In more recent years the captive cow Tara, who died on 28 April 1989 at 75 years, came close to breaking this record. One of Tara's long-standing companions in the camp at Mudumalai, the cow Godavari, also only just failed to break Peri's record.

Tara was captured on 15 April 1935 at Mudumalai and spent most of her life there. She was one of the most gentle of all elephants. Perhaps, as she spent the nights inside the jungle, her freedom hobbled, she sometimes met and socialized with her mother and sisters, or even introduced one of her dozen children to them. I wonder whether she ever yearned to be free like them.

In May 1982 M.A. Partha Sarathy came with a film crew to shoot a film on elephants for Indian television. They had only a few days to spare in my study area. By that time Meenakshi's clan had dispersed from the Araikadavu valley and it was difficult to get good sightings of the herds. So I took the team up to Kyatedevaragudi where Champaca's clan would still be coming to the pond. Even if no herds came, I knew that Biligiri would not disappoint them. He was a visitor almost round the year.

After driving around the jungles we took up position near the pond. I kept a sharp lookout while the crew relaxed, relating to them stories of all my wonderful sightings here to keep up their interest. After a couple of hours and no elephants, the crew was getting a bit impatient. Just as the cameraman began suggesting that we should perhaps pack up and move elsewhere, I noticed a flash of white moving among the trees. Biligiri climbed down the bund into the water, as I dramatically urged the crew to swing into action. Actually, there was no need for hurry as Biligiri was usually very relaxed.

Two months later Madhav visited me for a few days and we went around inspecting the damage to crops in the villages and trying to locate elephants. This time several herds appeared at Kyatedevaragudi. The herds remained in the now lush jungle until the end of August, when all of a sudden they made a bee-line south towards Punjur. As during the previous year the ripening *ragi* crop attracted the elephants' attention. This year, however, much of the crop had withered due to the poor rains and the elephants did not have the glorious feast they had had in 1981; though the patterns of raiding were similar to those of 1981, the monetary value of damaged crops was far less.

A solution had to be found to the problem of crop depredation. In November, the Asian Elephant Specialist Group arranged for a visit by Robert Piesse, an expert on electric fencing from Australia. I took him around my study villages to show the extent and circumstances of elephant incursion into cultivation. He explained to me the design of the high voltage electric fence that could 'knock the shit out of an elephant' coming into contact with the wire, yet not harm the animal in any way. He was confident that a combination of the fence and a beeper that gave a high frequency sound would be sufficient.

In December 1982 an international workshop on elephant management was held at Jaldapara Sanctuary in West Bengal state of eastern India. The workshop was very creatively organized with 'live' demonstrations of elephant capture, care, training, and control. There was even a simulated *kheddah* complete with a drive into a stockade. The workshop was attended by researchers and field managers from all over the world. This was a wonderful opportunity for me to exchange notes with so many other elephant people.

Cynthia Moss, who had been working for over a decade on African elephants, was there too and gave a graphic presentation of her Amboseli experiences. She described the intense greeting ceremonies elephants of the same family indulge in when they meet after even a temporary separation. She also outlined how elephant society is organized—into families, bond groups, and clans—the last level being obvious from the dry-season aggregation of elephants. I was particularly thrilled to receive this last bit of information; my own observations over the past two years having been vindicated.

I returned to Hasanur to find that the elephants faced a grim year ahead. In 1982 only about half the normal quantity of rain had fallen over the study area. The elephants had so far managed . . . but how would they cope with the dry spell that now loomed?

CHAPTER 7

Seasonal Rhythms

The great elephants, being happy with the smell of the screwpine flowers and disturbed by the sound of the waterfalls, roar, along with the peacocks, in the forest brooks.

Ramayana, Kishkindhakanda 28, 28. (c. 1000 BC)

Late in the afternoon of 25 January 1983 I was driving from Hasanur to Dimbam. As I began the climb up the first hill the shadows were lengthening in the valley to the east and a cool breeze was sweeping down the slope. The entire Mriga family of twelve elephants had just crossed the road and was moving down into the valley when I first noticed it. One of the Mriga sisters had a small U-shaped cut at the bottom of her right ear; her identity was unmistakable.

Another of the Mriga sisters, an older cow, had a calf at its heel. The calf was somewhat wet, indicating that it was no more than three or four hours old, and was covered with the soft brown hair so characteristic of new-born calves. It was a bit wobbly on its legs but managed to walk with the support of its mother's trunk. Flanking the calf and the mother was a cow much younger than the matriarch, probably an older sister of the calf. The other members of the herd were spread out along the dry hill slope searching for any green titbit that may have survived.

The new-born calf kept searching for its mother's nipples, sometimes at the wrong place—between her hind legs. Indeed, its search for its source of nourishment seemed to be somewhat at random between the front and hind legs, with about a fifty per cent chance of success. Whenever it was successful it suckled only briefly. By now the elephants had noticed me and naturally seemed nervous of my presence. The mother reassured her child by placing the tip of her trunk near the calf's mouth. Mother, child, and aunt, unsure of what lay ahead, stood huddled together as the sun sank behind the western hills.

After half an hour of observing them it was time for me to

leave. It was getting dark and I did not wish to disturb the elephants' peace. They would need to summon all their resources if they were to get through the coming months.

This was one of only two births I recorded in my sample of registered elephants during 1983. There would have been some more in unregistered groups, but there were certainly very few births in 1983. This was in a way natural because many calves had been born the previous two years and it was probably fortuitous that few elephant calves were destined to enter the world during the dry season of that year. They would have had to put up a real struggle in order to survive. Many of the mothers looked fairly emaciated and would not have had the energy to raise healthy children. The culprit was *El Niño* (meaning 'the child' in Spanish), a periodic heating of the equatorial Pacific Ocean. During 1982–3 *El Niño* came earlier than usual and stayed later. The result of this anomalous heating was catastrophic. Places as far apart as Central America, southern Africa, Australia, Indonesia and southern India were stricken with severe drought, while others reeled under floods.

As crops everywhere withered in the fields, cattle died by the millions in Australia, and elephants roamed far and wide in search of water in southern Africa. The elephants in my study area were also beginning to feel the pinch. The careful routines they had followed the previous two years now broke down. It was now that the knowledge accumulated by the matriarchs over the years became crucial for the very survival of the clans.

Before going on to describe how the normal routine of the elephants was upset during this deviant year, I will first describe those of a normal year.

Blue skies, warm days, and cool nights herald the new year. The leaves, which seemed green not too long before, begin to yellow and fall, crackling under foot in the jungle. The once verdant grassy carpets of the forest acquire a well-trodden look. The chital doe is emboldened to bring her fawn out of hiding, only for it to fall victim to the ferocity of the pack-hunting dhole. The sloth bear feasts on the ripe fruits of *Zizyphus jujuba* in the dry jungles. The *kutharis* of harvested *ragi* grow larger day by day in the fields. The farmers anxiously watch over the fruits of their labour, waiting for the right day in which to thresh the ripened grain. They have a few more sleepless nights because the bull elephants still lurk nearby, waiting for the cover of darkness in order to make a final grab at the harvested crop. The elephant herds now leave the vicinity of cultivation to move on to their dry season range, into the valleys, along the swamps, or up into the forested hills.

By February the blazing sun has sucked all the extra moisture from the tall grass swards. The forest floor is loaded with potential fuel, ready to ignite at the careless dropping of an unextinguished match or cigarette. In the teak forests, once the leaves are shed, the jungle turns a drab brown. Walking is tortuous during the day without a shield against the sun. And ticks, those tiny scourges all field biologists dread, invariably cause sleepless, itchy nights.

The tall swards of perennial grasses—*Themeda, Cymbopogon, Imperata*—are double-edged swords indeed. Their siliceous leaves are a deterrent even to a coarse feeder such as the elephant. Elephants frequent the gallery forests along perennial streams or, where a wide range of habitats is available, move into the patches of evergreen forest. They turn their attention to a variety of shrubs and trees or, if they are close to a swamp, to still-succulent grasses. In Mudumalai the Hambetta and Imbarhalla swamps are favourite haunts of elephants, much like the famous Lahugala tank in Sri Lanka. Since numerous elephant herds converge upon such nutritional hot spots, it is not uncommon to see a large assembly of anywhere between fifty and a hundred elephants in such places, sometimes spread out as distinct family or kin groups, or aggregating as a single clan.

Elephants also frequent the gallery forests and swamps because at this time of year water is usually only available in these places, though water may also be found in the small ponds dug for the use of cattle or, in sanctuaries, those made specifically for wildlife. Elephants do not need to drink every day, but they nevertheless require large quantities of water. An adult may drink 200 litres in a day.

The festival of *Shivarathiri* ('the night of Shiva') in March heralds a further change in weather. The nights are suddenly warmer. Temperatures which had been generally below 20°C at night now go beyond this mark, while during the day they shoot above 30°C in the shade.

Fiery necklaces adorn the hills in the late evenings. These fires sweep across the forest floor consuming the dense clumps of grasses, killing the young saplings, and reducing fallen trees to

A cow, possibly pregnant, secreting from her temporal gland.

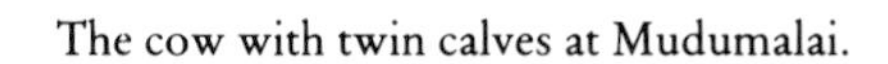

The cow with twin calves at Mudumalai.

A cow removes the foetal sac from her calf, which was born a few minutes earlier, and waves it about with her trunk.

The matriarch charges at the men carrying a barrel of water for the calf.

A calf suckling some six hours after its birth in captivity.

Divergent Tusks feeding on the bark of a young teak tree.

Sudha scans the panorama of shola forest and rolling grassland from Attikan estate.

A view of Hasanur, looking westwards from the road to Kollegal.

The *ragi* field of Basavappan at Hasanur damaged by Vinay.

Coconut trees damaged by a herd of elephants.

A grazier who sustained an injury to his left eye when he was flung by an elephant.

Dr K examines the skull of the raiding bull shot at Hasanur. Also in the picture are (clockwise) Ranger Premnath, District Forest Officer Ramachandran and Forester Narayanan

Two Sholaga children stand in front of their hut at Hasanur which had been pulled down by Vinay.

Two cows reassure each other with their trunks.

A calf reaches into its mother's mouth for a titbit.

ash. Drongoes, bee-eaters and rollers make merry, hawking the insects that rise with the smoke. An entire hill ablaze at night is a sight far more splendid than the glittering lights of the Ginza.

The fires are not caused by the behaviour of a careless smoker but by deliberate action. People living inside or along the periphery of the forest have for centuries regularly set fire to the forest for the purpose of cultivation and will continue to do so. Graziers set fires to promote a new flush of tender grass for their livestock; gatherers searching for tubers and fruits find their task much easier; people walking through the jungle to their homes can spot, well in advance, where there is an elephant bull or sloth bear lurking behind a bush. Poachers also set fires to cover their retreat, and people seemingly even set fire to a forest in vengeance against the government for curbing their access to its resources!

In the context of all these explanations as to why people set fire to forests, one fundamental point should not be overlooked—the direct adaptive value of fire-setting behaviour among primitive bands of hunter-gatherers or shifting cultivators. Imagine small clans of primitive people living in scattered settlements in the wilderness some 20,000 years ago (the use of fire by hominids actually goes back half a million years to *Homo erectus*). Their habitat is prone to fire, both from natural causes such as lightning strikes or the deliberate action of people. These clans knew the use and control of fire; they in fact used it to drive their animal prey to death and even to cook their food. If there had been no fires in their surroundings for a long time, the accumulated biomass or fuel would be prone to an intense conflagration which could easily develop into a forest canopy fire during an unusually dry spell. Imagine such a fire bearing down upon their tiny settlements at the speed of an express train, reducing to ashes everything in its path. One clan could thereby use fire to destroy an enemy clan; a natural fire would do likewise. Clans which periodically cleared their surroundings by setting relatively harmless ground fires would be insuring themselves against a fiery ordeal. Fire-setting behaviour in primitive man would have been favoured by natural selection. However, I do not claim that this is still a valid reason for fire-setting today.

In most jungles the network of roads and artificial fire-breaks prevent such ground fires from travelling for more than four or five kilometres at a time. Through a series of fires, a much larger area covering thousands of hectares may be burnt during a dry season. The deaths of elephants and other large mammals due to fire, however, is virtually unknown, although some small mammals, ground-nesting birds, reptiles, and a host of invertebrates surely perish. Fires are almost annual events in deciduous forests with tall grass but rarer in the short grass jungles.

If the orange glow of fires light up the hills at night, the trees are alive with even richer colours during the day. The flame of the forest *(Butea monosperma)* is loaded with hundreds of flowers, each papilionaceous corolla an orange lamp, inviting the birds to sup on its nectar. The coral tree *(Erythrina indica)* and the silk cotton *(Bombax ceiba)* put up an equally impressive display with their orange and red blooms. One could go blindfolded to these trees, guided by the cacophony of the mynahs—including hill mynahs, brahminy mynahs, and common mynahs—that flock to their flowers for nectar. A leafless Indian laburnum *(Cassia fistula)* overflowing with its yellow, drooping clusters of flowers, is a refreshing sight even on a hot, Indian day.

By now thunderclouds build up in the afternoons through local convection, sometimes vanishing as suddenly as they appear, at other times precipitating in short, sharp spells of rain. On a March afternoon the moisture evaporates as quickly as it would from a frying pan, leaving no trace of dampness by evening. It is only in April that the thundershowers begin to have a perceptible effect on the habitat. A new flush emerges from the burnt clumps of grasses whose roots have survived the fire; there are signs of a renewal of life. The air is filled with the intoxicating honey scent of *Terminalia bellirica* flowers. *Schleichera oleosa* flushes bright red leaves, making up for its lack of a floral display. The persistent '*brain fever! brain fever*!' call of the hawk-cuckoo haunts the evening hours.

By May strong winds drive more clouds across the hills. There is a minor peak in rainfall in areas such as Hasanur and the Sigur plateau which lie in the rain shadow belt. Water begins to trickle down the streams. If the rains are promising enough some farmers cultivate sorghum, maize, and gingili.

The elephants disperse from their dry season habitats, the swamps and the river valleys, spreading out into the deciduous forests with tall grasses. The burnt patches attract their attention; they eagerly consume the tender flush that emerges after the rains. Since the elephants are now spread over a wider area, the large groupings that existed during the dry season are no longer seen.

During a normal year the southwest monsoon reaches the tip of the Indian peninsula on the first day of June, advancing northwards to the Nilgiris and the Biligirirangans within a week. The terms southwest monsoon and northeast monsoon are apparently misnomers; the word monsoon derived from *mausim* (or season in Arabic) itself implies a pair of seasonal winds blowing in opposite directions. Whatever may be the etymology of the word, a nation of 880 million people waits in eager anticipation for that event known as the Indian monsoon. Though it sometimes plays truant, when it arrives it does so with a bang. The strong, moisture-laden winds drench the west coast and the slopes of the Western Ghats facing the Arabian Sea with copious rains. The eastern side of the ghats has to be satisfied with the leftovers. From a rainfall of 5000 mm. on the Mukurti ridge during the southwest monsoon season (June to September) the precipitation falls to only 250 mm. at Mangalapatti, a distance of only fifty kilometres away, lying sheltered in the Moyar valley. The better exposed regions to the east of the ghats, such as Wyanad, Mudumalai and the Biligirirangans receive a good proportion of their annual precipitation from the southwest monsoon.

By July, the rivers and streams are in full flow. The trees are now fully clothed in their new attire. Another wave of flowering begins across the different habitats—*Lagerstroemia microcarpa, Terminalia crenulata, Tectona grandis, Acacia suma* and a host of other trees. If 'large and bright' is the motto of the dry season flowers, then 'small and pale' is the rule of the wet season flowers. The peacock proudly shows off his gorgeous attire, strutting about in a frenzy to catch the attention of his lady love. The sloth bear gorges itself on the fruits of *Cordia*. The throat of the sambar turns an angry red, the so-called 'sore patch', the significance of which is still being debated. The whistling love call of the bull gaur pierces the humid air as he signals his presence to the cows in the area.

Elephants now start to visit the maize and sorghum fields, especially after the crops flower. Raiding by herds is sporadic but the bulls are more persistent. Before the retreat of the southwest monsoon in August or September, the farmers plant *ragi*, their staple food crop.

Come September, the rains intensify over the rain shadow regions—Hasanur, Talamalai, and Sigur. *Ganesh Chaturthi* (the festival of the elephant-headed god, Ganesha) is celebrated with great fanfare. No other city typifies the spirit of this day as much as Bombay. On an appointed day after the festival, large crowds, dancing in frenzy, snake their way through Bombay's lanes and thoroughfares, carrying Ganesha idols of all sizes and colours for the ritual submersion in water.

In the forests a different kind of collective movement takes place. The elephants regroup themselves into their respective clans to begin their trek from the hill forests, now covered with tall coarse grass, down to the more open jungle for feasting on the shrubs and the tasty short grasses.

October marks the beginning of the second inter-monsoon period. Although rainfall decreases over the western regions, the influence of the easterly winds actually increases rainfall to its annual peak over the eastern regions. Once the *ragi* flowers, the elephants descend upon the fields to satisfy their avid appetites for this crop. The night of *Deepavali* (the festival of lamps) resounds with the bursting of fire crackers all over the country. In elephant land the crackers are thrown at the raiders, usually with little success.

By November the tall grasses have grown to over ten feet in height in the moist deciduous forests. Very few elephants can now be seen here. They are spread out in the open jungles busy with their nocturnal excursions to the nearby crop fields. The pride of the peacock is deflated; he no more has his gorgeous train of feathers. Clusters of orange and red fruits adorn the *Terminalias*. The jungles are at the peak of their verdant attire.

The northeast monsoon establishes itself in December. The skies begin to clear and rainfall declines. The farmers keeping watch over the fields from their huts or tree-top machans move closer to the fires they kindle as the mercury dips below 10°C at night. Soon the harvesting of *ragi* and paddy begins, but the night vigils are not yet over. The elephants continue their raids on the *kutharis*. In the Biligirirangans the herds again come together into larger aggregations. The farmers look forward to the turn of the year, which will see the herds move away from the fields to their dry season range in the hills.

These cycles repeat year after year, generation after generation, varying subtly at times and more obviously at others.

During a normal year Meenakshi's clan would move east from the neighbourhood of Talamalai during January or February to the hills between Dimbam and Hasanur. From here the herds would go north to the Araikadavu valley before any rains fell and stay there until the end of April. The heavy pre-monsoon showers in May would see them disperse over a wider area into the surrounding hills and the open jungles further north near Punjur. They would persist until August or so and then turn south towards Talamalai.

In January 1983, with the drought worsening, the future was uncertain. Meenakshi's clan did move into the jungles near Dimbam as usual. Apart from Meenakshi's immediate family, I saw the Mriga family and High Head's family there. Tippu was also there, trailing the herds. Where the elephants went in February is a total mystery. Some elephants did come to the Araikadavu valley but did not stay long there. This was not surprising as the Araikadavu was totally dry save for a few pools of stagnant water on the rocks. The herds had to search for sub-soil water in the sandy bed.

One February morning, I sighted a herd moving north along the Araikadavu bed. I quickly went ahead to a spot where the steep embankment around a curve afforded a convenient and safe view. Soon the herd came slowly towards me and to my surprise halted just below the spot where I sat. Then I noticed that the sandy bed already had many pits excavated earlier by elephants. The adult and sub-adult cows now began deepening these pits further with their forefeet. As the water began to seep inside they sucked in the muddy liquid with their trunks, discarding a few trunkfuls before drinking. When resources are scarce even the thin line between altruism and selfishness disappears. Each animal fends for itself irrespective of considerations of kinship. When a juvenile eagerly came up to a hole for a drink of muddy water, a young cow, who could have been its sister, simply shoved it away.

This method of digging not only enables elephants to obtain some water from a dry habitat but also provides them with naturally filtered water, thus lessening the risk of infection from drinking contaminated stagnant water. Not only the elephants but the people of Hasanur also obtain their water in a similar fashion. My daily supply of two pots of water used to come from pits dug on the Araikadavu.

I never saw Meenakshi's family or any of the other identified families of her clan in the Araikadavu valley that year. Kali's family did come, but then I was never sure if it really had close bonds with the others in the clan. If the other herds had come in, as they very well could have, they also moved away quickly, and I have no idea where they went if they did. One perennial source of water was the Moyar river to the southwest, but to get there involved a difficult trek down the steep hills. Actually, it was not as difficult as it seemed for the elephants. The steepest hills in elephant country feature a network of precision-cut trails that would do a civil engineer proud. In fact, engineers have often used these elephant-created paths which gently embrace the steepest slopes, to align their roads. Another source of water was the Suvarnavati reservoir. During March and April, when over a hundred elephants would normally roam the Araikadavu valley, hardly ten elephants utilized the area on an average day. One result of this was that the pretty *Acacia suma* trees at Karapallam were spared from being attacked as in previous years.

If the *Acacia suma* escaped being devastated, other plants such as *Kydia calycina, Grewia tiliaefolia* and *Helicteres isora* received more than their usual share of damage. In April, when I went north to the Biligirirangans, I found an unusually large number of herds crowded into a small area near the township of BRT Betta, where the Somesvara pond still held sufficient water. These were mostly the herds of Champaca's clan, but herds from elsewhere may also have come there. The elephants would come to the pond in the evening for slaking their thirst and then turn to the plants in the nearby forest, breaking the stems, uprooting them or stripping the bark.

This brought home the enormous flexibility built into the lives of elephants—in their movements, foraging, reproduction and behaviour. They were not locked into a pre-programmed course, but could change with the circumstances, adjusting their lives to the vicissitudes of their environment. True, there was a conservative streak to their seasonal movements, but the very factor responsible for this conservatism—the legendary memory of the matriarchs—also contributed to the flexibility that steered the family through difficult times.

My observations of Champaca's clan exemplified this point. In January, Champaca's clan moved from the scrub jungle around Punjur and Kolipalya northwards into the denser hill forests of the Biligirirangans. Here the herds seemed to largely move through the semi-evergreen patches and gallery forests in the hills. With the beginning of the pre-monsoon showers in April they certainly ranged more widely over the deciduous forests, often visiting the pond at Kyatedevaragudi. This pattern continued through the early rainy season until September when they began their southward exodus down the hills to the open jungles. This remained their home until the harvest of the *ragi* crop at Punjur–Kolipalya in December.

Reading G. P. Sanderson's book, I came across the following observation:

> In the dry months—that is, from January to April, when no rain falls, the herds seek the neighbourhood of considerable streams and shady forests. About June, after the first showers, they emerge to roam and feed on the young grass. By July or August this grass in hill tracts becomes long and coarse . . . elephants then descend now and again to the lower jungles, where the grass is not so far advanced . . . the herds invariably left heavy jungle about October for more open and dry country. About December, when the jungles become dry, and fodder is scarce, all the herds leave the low country, and are seldom seen out of the hills or heavy forests until the next rains.
>
> In the jungles around Poonjoor [i.e. Punjur] when there is plenty of rain, game of all descriptions, from elephants downwards is abundant.

The pattern of movement that Sanderson had observed in the Biligirirangans during the nineteenth century was exactly what I observed for Champaca's clan a hundred years later. Champaca's great-grandmother may well have then been the grandmatriarch, leading her kin up the hill slopes and down into the valleys, through good and bad years, with uncanny judgement in the struggle for survival. The wisdom of the great matriarchs is passed down through the generations in this long-lived mammal.

One of the aims of my study was to determine the sizes of the home ranges of the elephants. This information is important in designing protected areas for elephant conservation. If an elephant clan needed hundreds of square kilometres of forest it is no use in setting aside small patches of tens of square kilometres for them. I had to be satisfied with crude estimates; I did not have facilities such as radio-telemetry during my study.

By plotting my sightings of identified herds on a map I calculated that the minimum home range size of Meenakshi's clan was 115 square kilometres and that of Champaca's clan was 105 square kilometres. The actual home ranges of both the clans would have been higher, as they would have moved outside this range to peripheral areas where I would have missed sighting them. Assuming that a clan utilized the entire area of a habitat zone they entered I speculated that their home ranges were about 200 to 300 square kilometres. Studies in Africa have shown that elephants make occasional sudden movements into new areas for short periods of time. There was no way in which I could detect such movements.

The movements of other elephant clans in the study area were not very clear. One large clan (which may have been actually a sub-population consisting of more than one clan) moved into the Moyar river valley during October and persisted there until the end of February. Some of these herds certainly came from and retreated into the Nilgiri hills, the Sigur plateau and Mudumalai. Another clan was seen during December to February to the west of the main Biligirirangan hill range, in the vicinity of Gaddesal, but I did not register any of the families or know where they went the rest of the time.

The bulls seem to have their own distinct movement pattern. They did, of course, spend time with the family herds, but also lurked in the vicinity of villages on their own, when no herds were nearby. Vinay spent most of his time near Hasanur, moving west to Talamalai in January or February, where he associated with family herds, among them those of Meenakshi's clan. He invariably returned to Hasanur by May. His home range covered at least 170 square kilometres. The two locations he was seen farthest apart were twenty-one kilometres from each other.

Cross Tusks similarly moved at least twenty-one kilometres between Kyatedevargudi and Karapallam. I could not estimate his home range size because I had no sightings of him outside of a narrow belt between these two places. This was one of the frustrations of having to rely only on chance sightings of known elephants. Certain cows and bulls were frequently seen at the same spot, such as a favourite pond, but I could hardly consider these as resightings for statistical purposes. Tippu used to appear promptly in the jungles to the south of Hasanur during February to March every year. Makhna was seen regularly between February and July, always in the Araikadavu valley. Where they spent the rest of the time was a mystery to me. The bull with the largest home range that I recorded was Divergent Tusks. After spending the most part of the wet season during May to November in Mudumalai or Bandipur, he used to move east into the Sigur plateau. On 27 December 1982 I saw him unexpectedly in the Moyar river valley near the Lower Bhavani Reservoir, a distance of fifty-two kilometres from Kargudi. The valley was now teeming with elephant herds and Divergent Tusks no doubt was following them to mate with oestrous cows. His range encompassed an area of 320 square kilometres.

After a pubertal bull leaves its family it may initially confine itself to a small range, sticking close to its natal clan. Young bulls like Biligiri seemed to exemplify this pattern. As a bull grows and matures socially it may expand its movements to cover the home ranges of several clans. This would enable it to mate with several cows. Vinay and Divergent Tusks were certainly bulls in their prime. Such bulls may make transitory movements through new areas before settling down in a place. Even with several years of familiarity with an area, at times one comes across bulls never seen before. A bull's range may once again shrink during old age when it may seek the security of water, fodder, and shade. Old Dodda Sampige seemed to typify this behaviour. I never saw him outside a small area in the northern Biligirirangans. Much of this is of course sheer speculation on my part. I can only hope that a well-organized radio-tracking study of this aspect may one day be done.

Fortunately for the elephants, after the long dry period, the pre-monsoon showers came on time in late April 1983 and by May the landscape was once again on its way to regaining its mantle of green. The elephants had come through the drought with exceptional resilience. Apart from one calf that died near BRT Betta, no other death could be directly attributed to the drought. The death rate during the preceding few months was hardly different from the normal rate. It was heartening to see that the elephants could cope with the stress, but viewed from another angle this raised new questions. Were the elephants being artificially maintained at high densities with the assistance of man-made water sources, such as ponds and reservoirs? If more elephants did not die even during a drought, would we see a population boom that may eventually damage the vegetation and cause a massive crash in elephant numbers similar to the African experience at Tsavo?

The growth of fresh nutritious grass and leaves with the onset of the rains would have stimulated the reproductive cycles in the elephants. By this time my field work was gradually tapering off as I had to get down to the business of compiling my data. I could not make any observations of reproductive behaviour that wet season, but two years later I noticed a baby boom in the nearby Nilgiris. I have no evidence that the same was true for the Biligirirangans but presume this must have been the case as the two regions are part of the same elephant range. This was also the first time that I got a hint of seasonality in breeding. A large number of mature cows seem to have given birth between July and October 1985.

Seasonality in mating during the wet season, and giving birth after a gestation of twenty to twenty-two months, just before the onset of rains, would be very advantageous. The lactating mothers would be assured of sufficient food for nourishing themselves and suckling their children. In real elephant populations, however, such patterns may or may not be seen. An interplay of variation in rainfall seasonality and oestrous cycles in cows could obscure any clear pattern in reproductive seasonality. In southern India, I have seen very young calves throughout the year, including the dry months. Among captive elephants, in fact, most babies are born between December and February at the beginning of the dry season.

In order to understand the dynamics of the wild elephant population, I had to find out, among other things, how fertile it was—that is, at what age do the females begin breeding, how often do they produce children, and when do they stop breeding? From my registered families I saw that hardly any cow less than fifteen years of age had a calf, while most of the ones above twenty years had one or more; the cows thus seemed to give birth to their first calf between the age of fifteen and twenty years. I therefore assumed the average age at first calving to be 17.5 years. This means that they became sexually mature at the age of fifteen years or thereabouts.

These figures were supported by records of 260 calves born to captive elephants in southern India. The earliest age at which a cow gave birth was at 13.3 years, that of Meenakshi, who was herself born in captivity during the last century. All other captive elephants have calved only after the age of fifteen. I realize that these figures may be different for other populations. In particular, the Sri Lankan elephants seem to calve at an earlier age.

The next question to answer was how often does a cow give birth or, expressing it in different terms, what proportion of mature cows give birth during a year. From my registered families I drew out all mothers who had at least two children below nine years old (aging becomes rather imprecise for this purpose beyond this limit). The average difference in their ages was 4.6 years. Looking at it from another angle, I found that 63 per cent of mature cows gave birth during 1981–3. This gives an average calving interval of 4.8 years, which matches with the earlier figure. But this was only part of the story. When the annual birth rates were considered I saw that they fluctuated widely. For instance, 22 per cent of cows gave birth during 1981 and 33 per cent during 1982, but only 8 per cent did so during 1983.

When I went through the records of eighty-eight elephants captured by *kheddah* in 1968 in the Kakankote jungles, I noticed a similar cyclic pattern. Eighteen of the thirty mature cows had given birth within the span of one year, an extremely high rate for that year. There were few calves aged one to three years, but many calves which were three to four year old, indicating that there had been another boom in births three years earlier.

Such inter-annual cycles in elephant births have been earlier recorded for African elephants by Richard Laws and Ian Douglas-Hamilton. But this was the first clear demonstration of birth cycles in Asian elephants. It is easy to see how a fluctuating pattern in births can arise if we take a year as our unit. The gestation period of elephants is nearly two years. If a large number of female elephants happen to conceive during a year of exceptionally favourable environmental conditions, they would naturally not be able to conceive the following year while still carrying the unborn, or even the second following year when they are still suckling their calves and hence have not resumed their oestrous cycles. The earliest they could conceive again would be three years following the previous conception. This would give rise to a cyclic pattern in births with peaks every three years. In nature the pattern would never be so regular over a long time because a dozen other factors would superimpose upon this their own peculiar patterns. Variation in rainfall is one obvious factor that could influence births.

I tried to correlate the birth rate during a particular year with the rainfall two years prior to it but failed to see any clear pattern, although I suspect that its influence lurked somewhere. If anything, this underscores the need to monitor elephant populations for many more years before we can hope to decipher the intricate details of their demography and dynamics.

Finally, I estimated the age at which elephants stop breeding. An exact figure was not terribly important because there would be very few old cows, whose contribution of children to the population would be negligible. Some of the oldest matriarchs in my study area, including Meenakshi, who exceeded fifty years of age were still reproductively active. Among captive elephants, the cow Tara gave birth to her last calf when she was sixty-two years old.

Tara, in fact, must have been one of the most prolific breeders among captive elephants. From the time of her capture in 1935 until her last calving in 1976 she went through twelve pregnancies and produced thirteen calves (including one pair of twins, one of which died soon after birth). Twins among elephant calves are quite rare. Out of 260 parturitions among captive elephants there have been just three instances of twins. Apart from Tara's twins, other twin calves born to Devaki and Valli survived. In the wild, it would be uncommon for both the twins to survive as the mother may not have sufficient milk for them. Only once have I seen twin calves among wild elephants; this was at Mudumalai.

Although the elephants got through the 1982–3 drought without any scars, they could easily have succumbed to a variety of problems. Afflictions of the digestive system are common in elephants. Colic is a particular irritant. When sufficient water is not available, the animal may become constipated or, worse still, suffer from intussusception or telescoping of the intestines. By drinking contaminated water they may be infected by tapeworms, roundworms and hookworms. In severe cases this can result in diarrhoea.

Some years later I watched a elephant cow suffering from diarrhoea in the throes of death. This was in May 1987, another drought year in southern India. Out of exhaustion the elephant had settled down in a stagnant pool of water in the Kekkanhalla stream, along the border of Mudumalai and Bandipur. It made numerous efforts to stand up, but collapsed each time. Occasionally it would reach out for a clump of grass or a trunkful of bamboo leaves along the stream bank. By night it seemed that its hours were numbered.

The next morning a *koonkie* elephant from the Theppakadu camp was marched to the spot. The cow was still alive, thrashing the water desperately and trying to get up. After an hour it stood up to our surprise and began to walk slowly. It got out of the pool and hobbled up towards us. There was faint hope that it might survive, but nothing much could be done to help the animal. By afternoon it collapsed again after walking over 200 metres. This time there was hardly any movement but the end had not come yet. The cow managed to get up during the night and walk another half a kilometre before it died the next day.

Elephants are also susceptible to various communicable diseases, including anthrax, tetanus, tuberculosis, pox, and rabies. Apart from a few cases of anthrax none of the other diseases is known to have actually killed any wild elephant in southern India in recent years. Elephants also succumb to diseases of the heart, the circulatory system, and the lungs. A bull that died in Satyamangalam had a tumor the size of a football in one of its lungs.

Many elephants meet an accidental death. Normally sure-footed, elephants are known to have fallen down steep hill slopes; calves have got stuck in rock crevices and have starved to death; those venturing to drink from reservoirs have been bogged down in the soft mud along the shores. In the Anamalai hills of southern India, elephants have been carried away by the swift current in canals connecting a reservoir with a power house and stranded at an entry into a tunnel. Some have been rescued with great difficulty by using tame elephants, while others have simply drowned or died of exhaustion.

Dead elephants are examined by a veterinarian to ascertain the cause of death. The post-mortem time examination of an elephant is an unforgettable experience. A three-ton cow or a five-ton bull is a formidable object whether alive or dead. Dr K has developed the investigation into a fine art with over two hundred elephants in his bag. After he cuts open the elephant he literally gets into it, turning over the liver, lungs or the heart, pulling out what seems like miles of steaming intestine, examining the brain, and collecting sample tissues. After the operations are over, he cleans his hands and calmly eats his lunch, a purely vegetarian meal, while most of his companions who had refused to come near the stinking carcass dine on meat.

The post-mortem reports were invaluable for me in order to analyse mortality patterns. The death rate in female elephants was relatively low, being less than two per cent per year between the ages of five and forty. Male elephants on the other hand, suffered a ten to fifteen per cent mortality per year. For both sexes it was difficult to get reliable estimates of death rates in juveniles because their carcasses might have gone unnoticed. Male calves have twice as much risk of dying as do female calves during the first few years of life.

There was another important difference in the mortality patterns of male and female elephants. Most of the females died a natural death, while most of the male deaths could be directly attributed to man.

Since the bull elephants raid crop fields more often, they also have a higher risk of dying there. In November 1982, the Karapallam Bull had fallen in a farm at Hasanur. He had negotiated a two metre deep trench and entered the farm only to come in contact with an electrified wire. One out of ten bulls die due to electrocution or being shot while raiding crops. Chemical mixtures which are meant to explode on compression are hidden inside jack fruits or balls of cooked millets and left in the fields. Although these are usually meant to deter wild pigs, many a time an elephant has picked up one of these and consumed it, with fatal consequences.

In July 1983 at Nagarhole, in the company of Ullas and Chinnappa, I came across an adult bull with large wounds to its trunk, near its mouth and tusks. It had undoubtedly been tempted by one of these lethal food packages in one of the estates adjoining the park. In spite of fresh grass being available in plenty, it was in poor shape as it could no longer feed easily.

The sun had already set and I had only a 64 ISO transparency film in my camera. Nevertheless, I wanted to get some pictures of the wounded bull for the record. Chinnappa was confident that we could get close to it and so the three of us crept up to a tree, watching it painfully pluck small clumps of tender grass and place them in its mouth. Steadying my camera with its 200 mm. lens against a tree trunk, I exposed a few frames at 1/30 second shutter speed. Considering the poor light, I was pleasantly surprised at the results. We could not make out if the bull had any injury inside its mouth. Two months later Chinnappa found the bull still alive with its wounds partly healed.

Most elephants are not so fortunate. One four year old male had its jaw fractured and a hole blown through the roof of its mouth. It died within a day inside a tea estate in the Nilgiris. Actually, such incidents are fortunately not very common. Most of the farmers are relatively tolerant of damage to their crops. When an elephant is injured or killed it is usually by one of the richer farmers.

A far more insidious menace threatening the lives of the male elephants is the lust for white gold. A wave of ivory poaching was sweeping through the southern Indian states, felling the old bulls with six foot tusks and the young ones with one-foot tusks alike. Seven out of every ten bulls that died were mown down by the poacher's gun. The 1980s was the decade of the ivory poacher. The African elephant population crashed to alarmingly low numbers under the merciless wave of the ivory seekers. The Asian elephant has so far managed to hold its ground, however, as the females do not carry tusks, but most of the majestic tuskers were gone by the end of the decade.

To tell the story of poaching I have to go back in time, to the early days of my study.

CHAPTER 8

The Ivory Hunters

> Thus, O king, having reached a hundred and twenty years, and having performed many deeds, the elephant goes to heaven.
>
> Nilakantha in *Matanga-Lila*

Appu could sense that something was wrong during the previous few days. His mother and his aunts were extremely restless, frequently sounding an alarm, herding them together, and running to and fro with their trunks in the air. Things had been this way since they had come across the lifeless remains of his distant cousin, a young tusker of his own age with whom he had often played and fought when their families had met. Now the cow elephants were again agitated, with their trunks raised and scraping the ground with their feet and throwing mud over their backs. Meenakshi gave a call for them to bunch and all the little ones soon got into the tight cordon formed by the elders. Appu, was a little slow to respond; after all, he was five years old and had already begun to explore the world on his own at times.

It was then he saw the dogs, sleek brown hounds, running down the hill slope. Instinctively, he ran towards the rest of the herd. In a few moments the pack had surrounded them, barking and attempting to nip at their legs. Meenakshi rushed forward at one of the dogs but it easily dodged her. He felt a sharp pain in his left hind leg and wheeled around roaring loudly to see one dog bounding away. The war of nerves continued for sometime, with Meenakshi charging and trumpeting, the dogs nimbly dodging her, until one of his aunts, who was of a somewhat nervous disposition, panicked and broke away, followed by some of her youngsters. The herd then scattered and suddenly Appu found himself alone, surrounded by the dogs.

In fright, he crashed down the slope into the dry nullah where the bamboo thickets afforded some cover, only to hear a loud bang and see the sand kicked up in front of him. He ran as fast as he could as two more shots rang out. By now he was in fairly dense cover but he kept going for another hour before he realized that the dogs were not on his trail. It was getting dark and he was frightened, but at least he was still alive. It took him three days to find his family again.

The above narration is fictitious, but a plausible one based on accounts I have heard about how ivory poachers once operated in the Bargur hills. Such ingenious methods of using trained dogs to separate young tuskers from the herd, before shooting them down, are not practised today. But ivory poaching continues to be lucrative in southern India. Sophisticated automatic rifles have replaced the traditional muzzle-loading guns.

When I began my study, the first cases of poaching I knew of were in the Biligirirangans of Karnataka. In March 1981 one bull was poached in the Punjur Range, followed by two more in May. Then, in October, one of the crop raiders was shot at Hasanur and its tusks removed. This was, as I have mentioned earlier, only opportunistic poaching and not the work of professional poachers. The same year poachers killed at least ten tuskers in the nearby Kollegal and Mandya divisions.

The worst was still to come during 1982. Over a dozen elephants were shot in the Satyamangalam division. The exact number was not known because some over-anxious forest guards burnt the carcasses, fearing the wrath of their superiors. In fact, poached carcasses were burnt not only there but also in many other forest divisions. This only exacerbated the situation. The poachers invariably came to know of this and were emboldened to kill more elephants. A carcass certified by a veterinarian as poached was needed to convict those responsible in a court of law and having burnt it, this evidence was simply not there, even if the poacher was apprehended.

In the forests, traversed daily by numerous graziers, even the charred bones of an elephant cannot always be hidden. Rumours of poached elephants began to slowly circulate among the villagers. With the help of Setty and our contacts in various villages, I was able to examine many such instances of poaching. This data was important to me, as I had to get a realistic estimate of the death rates in both sexes in order to know what impact these would have on the elephant population as a whole.

Following my investigations, it became rather risky to walk in the jungles, especially to visit the sites of poached elephants because the poachers had sent out a warning through the grapevine that they would shoot anyone they saw in uniform, meaning any staff member of the forest department. In my green field clothes I could have been easily mistaken for an official. I saw a poacher only once but he fled as he was alone and was probably only a scout. Setty was nervous of taking me to certain places, but he nevertheless did unearth a number of poaching incidents for me.

One of the Rangers in Satyamangalam, the formidable V. Chidambaram, was tracking down the poachers in earnest. It seemed that a small group of four or five men from the hamlet Irayanpura in the village Kolipalya, situated in Karnataka but adjoining the border with Tamil Nadu, was responsible. Their game was to shoot the elephants in Tamil Nadu and then retreat into Karnataka.

Such a pattern was common in the south; poachers from one state would operate in an adjacent state and then quickly return to their home state, where they would be relatively immune to arrest. Each state blamed the other for its problems, but the truth was that each of them had its own share of poachers. To catch an offender from one state the co-operation of local officials from the other was needed, but this was not always forthcoming, especially from the police. The police were concerned with the murder of people; the murder of an elephant meant little to them. The forest departments did have better co-ordination, but this usually depended on the personal rapport between the concerned officials.

Chidambaram vowed to put an end to the killing of elephants and the illicit removal of sandalwood from his area, a mission for which he was to ultimately pay for with his life—but that is getting ahead of the story. The District Forest Officer of Satyamangalam, M. Ramachandran, and Chidambaram confided in me their strategy. A trap was to be laid for the poachers by approaching them as purchasers of ivory. Through a friend, Chidambaram established contact with the person who acted as a middleman or courier for the tusks.

One day in July 1982 Chidambaram triumphantly flourished an audio tape and announced, 'I have a recording of a conversation with the elephant poacher.' As we played the tape we gathered the details of how elephants were shot, the tusks removed, and then taken out of the forest. We learnt that the poachers had also poisoned the carcass of a tiger's kill and were searching for the animal in order to acquire its skin.

A deal was stuck with the poachers for the purchase of a pair of tusks. The tusks and the money were to be exchanged during night at a predetermined spot along the road from Kolipalya to Talavadi. One night in August, Chidambaram accompanied by his staff and the police in plain clothes kept the rendezvous, posing as purchasers of the tusks. The poachers came up with a small pair of tusks weighing about a kilogram each. These would have come from a young elephant hardly six years of age. The 'purchasers' started a dispute over the weight of the tusks and at the opportune moment pounced upon their prey. In the ensuing mêlée one of the poachers escaped into the jungle.

This gang's arrest did bring down the incidence of poaching of tuskers in Satyamangalam. But this poaching was only the tip of the iceberg. A more widespread network of elephant poachers operated in the south. A village called Gopinatham in the Kollegal Division of Karnataka was the hub of the poaching operations. The poachers from here were the most ruthless and best organized. They operated not only in the nearby Madeshwara and Bargur hills but over a wide area extending into the forests of Hosur, Satyamangalam, Sigur, Mudumalai, the Biligirirangans, and Bandipur. Their exact relationship with the smaller groups was not clear, but there was a common pattern to the organization of the illegal ivory trade throughout the region.

The men who actually killed the elephants were local villagers. They roamed the jungles in groups, carrying firearms and food, shifting their camp regularly, as they searched for tuskers. Once an elephant was down its tusks were removed as quickly as possible. Some simply hacked off the exposed portions with an axe; others used quick-lime or acid to soften the tissues at the base so that the entire tusk could be pulled out. The tusks were taken out of the forest on foot or buried there until a purchaser was found.

The kingpin in the illegal trade in tusks was the middleman. This person usually came from a nearby town and it was he who organized the operations. He provided the weapons and took charge of the tusks for their subsequent sale to ivory dealers. Most of the ivory was sold to dealers in Kerala. It was common knowledge that one Sevi Gounder, who came from a town in Tamil Nadu organized the poachers of Gopinatham, but he was never effectively convicted. Ivory dealers who purchased poached tusks were usually licensed to sell imported African ivory. Since it was not possible to distinguish between African and Asian elephant ivories, the dealers could easily pass off the illegal tusks as those legally imported in case their stocks were checked by the authorities.

After their initial success, the forest departments in the southern states intensified their patrolling, but looking for poachers in the jungle is akin to searching for needles in a haystack, though by chance a few poachers may be apprehended. Each guard has to patrol anywhere from 10 to 100 square kilometres of forest, usually unarmed. It is another matter that in some regions the staff hardly venture into the forest; even in those where patrolling is well-organized it is nevertheless difficult to eliminate poaching. When a poached elephant is placed on record there is usually an uproar locally. The forest staff can even be suspended! In one state, the cost of the tusks was recovered from the Ranger's salary! No wonder that here there are hardly any elephants poached, if one is to go by official records.

The media are constantly on the lookout for lapses by officialdom, and poached elephants made a good story. I was often approached by journalists for news about poaching. They were usually more interested in this than in any other aspect of my research. I gave them the broad details and urged them to get the specifics from the officials of the forest department because I believed that such a dialogue was in the best interests of finding a solution to the problem. Some wrote balanced accounts of the poaching problem, while others were more sensational. Ultimately, the flood of articles in the press increased public awareness and spurred a more open debate as to how to tackle the issue.

Many bulls also die of other human-related causes. The injured bull elephant I saw at Nagarhole, a well-protected park, in July 1983 was typical of such incidents. The males were suffering a much higher mortality than were the females and it was imperative to know what effect this would have on the elephant population.

After my visit to Nagarhole, it was time for me to shift my base back to my institute in Bangalore. I had been at Hasanur for nearly two and a half years during the intensive phase of my field-work. The data I had gathered had to be interpreted and written up. Hopefully, this data would also provide the scientific basis for the future management of elephants.

Once I had shifted to Bangalore I only made brief trips to Hasanur and to Mudumalai. In October, during a visit to Mudumalai, I was shocked to learn that Divergent Tusks had been gunned down by poachers near Masinagudi. Selvam and I never actually saw his carcass but Selvam's tracker Krishna confirmed the story. At first we had lingering hopes that the poached elephant was not Divergent Tusks and that he had gone off somewhere else and would return one day, but we never saw him again. Sadly, Krishna was right. However detached a scientist may try to be about his subject it is not always possible to divorce oneself from one's feelings especially when the subject happens to be someone like Divergent Tusks. It was difficult for both of us to imagine that we would never see this gentle giant among the bamboo fronds along the Moyar, patiently pulling down the tall culms, posing for the most amateurish camera buff, and looking at that impatient honking motorist as if to say, 'Why are you going through life in such a hurry?' I can hardly bare to imagine what a ruthless poacher would have done—to go right up to him and put a bullet into his brain, betraying the trust Divergent Tusks had in people.

It seemed as though the Gopinatham gang changed their focus to Mudumalai and Sigur, where some very fine tuskers still remained. Many other bulls also fell here to the poachers before they once again shifted their attention to other areas. 1983 thus closed on a depressing note; it was uncertain what course the poaching wave would take.

The next year personally dawned more brightly for me. In January I married Sudha Swaminathan whose family had been known to mine since I was a boy. A month later we were off to Hasanur to see my elephants. For Sudha this was a totally new experience. She had never visited a real forest, much less had an encounter with wild elephants, but she gamely went along with the trip. At the Karapallam pond we saw Meenakshi with her son Appu, now about five years old, and another calf that had been born the previous year. This was a birth record that I had missed. From Hasanur we drove up to the Biligirirangans. At Kyatedevaragudi young Biligiri made his customary appearance, but the trip was too brief to try and locate the other elephants I knew.

Back at the institute I had to get down to work on my data. After leading a *junglee* life-style for several years it was very painful to sit down at a desk and write for more than ten minutes at a time. Every now and again I would get up and restlessly walk to the window to look at the trees outside before resuming the job of tabulating numbers, cross-checking notes, or reading a technical paper.

One of the first things I did was to compile the information I obtained on the illegal ivory trade in southern India and how this related to the international market. It turned out that up to 150 male elephants were killed each year by man, roughly in equal numbers in each of the three states, Karnataka, Kerala, and Tamil Nadu. Some of these were killed in defence of crops. A few of the elephants that were shot had their tusks intact; probably the poachers were not able to retrieve them for some reason. The tusks were stolen from at least 100 elephants annually during the late 1970s and early 1980s. This meant a harvest of about 190 tusks by poachers, assuming that some elephants were one-tusked.

My next task was to estimate the mean weight of a poached tusk. Poachers shot tuskers as young as five years of age; at this age the elephant would have yielded a pair of tusks weighing less than two kilograms. By comparing the proportion of males killed in different age classes with their occurrence in the wild population, I saw that males in the five to ten year age class were killed in a lower proportion than their availability. Those above ten years of age were shot in roughly equal proportion as they would have been encountered in the population. I calculated the average weight of a poached tusk to be 9.5 kilograms. Thus, the 190 poached tusks contributed 1800 kilograms (1.8 tons) of ivory to the illegal trade. The value of this ivory at 1982 prices was the equivalent of US $ 270,000 in the Indian market.

Compared to the enormous volume of international trade in African elephant ivory, estimated at 680 tons in 1980, the supply of 1.8 tons of poached Asian elephant ivory from southern India was a pittance. Why then were elephants poached here?

There are several reasons for this. India has the largest number of ivory carvers in the world. A survey in 1978 by

Esmond Martin, an expert on the international trade in animal products, gave a tally of 7200 carvers in the country, of whom about half hailed from the south. Historically, the supply of ivory from within the country has never been sufficient for the carvers. Ivory from Ethiopia was being imported as far back as the sixth century BC. During the late nineteenth century, and again prior to Indian independence in 1947, an average of about 250 tons of African ivory was being imported annually. After 1947 this declined sharply, with only five to fifteen tons being imported yearly during the 1980s.

A major reason for the decline was the high customs duty of 140 per cent imposed on ivory imports. Although the cost of raw ivory in the international market was only half that of local legal ivory sold by the forest department, the duties meant a higher landed cost. The carvers could purchase ivory from poached Indian elephants at a cheaper rate than by importing African ivory.

The forest departments in the south together sold only about one ton of ivory each year. Of the imported African ivory, the major share went to carvers in the north who worked with machines and thus had a higher output. The carvers in the south worked with their hands. They carved less ivory but their finished products were of superior quality and value. Of the approximately ten tons of ivory imported annually, less than two tons reached the south. Added to this, the forest departments in the south stopped the sale of tusks from their stocks during the early 1980s in an attempt to impose a total ban on the trade in Indian elephant ivory. Thus the supply of 1.8 tons of ivory from poachers was significant for the local trade.

It is well known that most of the African tusks went to Hong Kong and eventually to Japan, where people invest in raw ivory much like they do in gold or real estate, as a hedge against inflation. The Arab countries also began to buy large quantities of African ivory. There are reports that some of this has been smuggled into India, carved, and then taken back. The trade in ivory is thus a very complex affair.

Armed with my estimates of mortality and fertility of the elephant population during 1981–3, and knowing its age composition, I set about modelling its dynamics using a computer. The results were somewhat surprising. The elephant population was either stable or increasing at a slow rate of up to two per cent per year. The relatively low death rate among females was a buffer against a decline in population numbers. True, the selective poaching of males had caused a distorted sex ratio of one adult male for every five adult females, but this had not yet had any adverse effect on the population's growth.

One possible reason for this could be that the sex ratio for the purposes of reproduction in the population would have been much less unequal, perhaps one male for every two or three females. This would arise because a substantial number of cows would be either pregnant or suckling their calves and so would not enter oestrus. Thus the available cows could have all been mated by the bulls, though the latter were fewer in number.

Another possibility is that the younger adult bulls, those fifteen to twenty years old, would have been able to mate in the absence of the larger bulls. Although the older cows may reject advances from a young bull, the same may not be true of the younger cows. In any case, a 1: 5 sex ratio is not cause for any undue worry.

This is not to say that everything was going well for the population. The computer simulations projected a further decline of up to 40 per cent in the proportion of adult males in the total population over the next five to seven years, even if poaching were to be controlled, irrespective of whether the population as a whole was stable or increasing slowly. This also meant a greater inequality in the numbers of adult males and females; a ratio of about 1: 9. What consequence this will have on the population's fertility only future work will give the answer to.

My thesis on the ecology of the elephant and its interactions with people was completed and submitted and in December 1985 I was formally awarded a doctorate for the work. One of my thesis referees, Tony Sinclair, well-known for his work on the African buffalo, recommended that my work be published as a book. So it was soon back to the desk for me. Sudha had by now resigned herself to my endless hours of writing.

Our daughter Gitanjali arrived the next year in February. A few days later I joined the faculty of the Indian Institute of Science. The ecology programme at the institute had now fledged into a full department and provided opportunities for a variety of new research. In 1986 I was busy as a new father and in organizing new research programmes in the Nilgiris. The country's first biosphere reserve had finally been established there. Madhav and I had done much of the spade work for creating this reserve, including providing the design and management guidelines. Our centre was thus entrusted with a major responsibility for research in this region.

The poaching wave meanwhile continued unabated. In fighting this menace, the state of Karnataka took the lead. Its Chief Wildlife Warden, M. K. Appayya, saw to it that most if not all cases of ivory poaching were at least put on record. This motivated the staff to fight the poachers, without the fear of punishment if an elephant was killed in spite of their efforts. Karnataka was also the first to organize a system of regular patrolling using a wireless network for rapid communication. In the Biligirirangans, P. Srinivas, the young Deputy Conservator of Chamarajanagar, assisted by C. Srinivasan, began compiling a portfolio of poachers and tracking them down in earnest.

At Nagarhole, Ranger K. M. Chinnappa went hammer and tongs against the poachers. If Chidambaram was dark and hefty, Chinnappa was fair and lanky. Both of them were unmatched in their zeal to crush the poachers. A dozen Chidambarams and Chinnappas is all that is needed to give the poachers a run for their money. But, unfortunately, there are too few of them.

Now a more aggressive brand of poacher has arisen. Certainly the poachers are much better armed than the forest guards. They have even acquired automatic weapons. I suspect that the inflow of arms into southern India during the ethnic conflict in neighbouring Sri Lanka had a role to play in this. Poaching cannot be eliminated merely through combing the jungles. A good intelligence network is needed to identify and nab the crucial men. There also has to be better co-ordination between the authorities in the different states.

The aging Sevi Gounder seemed to have retired from the business, his place taken by Veerappan, who once worked for him, and who was wanted on several murder charges. In October 1986 the Karnataka police made a prize catch—that of Veerappan himself, on the fashionable Brigade Road in Bangalore—while they were on a routine interrogation of suspected trouble makers before a scheduled meet of heads of government from South Asian countries. Veerappan confessed that he was not after any head of government, only that of an elephant. He was taken to the Biligirirangans, where he was lodged at Budipadaga not far from the site of the first successful *kheddah*. Here, the outlaw was photographed for the first and last time by the authorities. The same night Veerappan escaped, allegedly with the connivance of the police, a lapse which many, including the police, were to later bitterly regret.

In late 1986 there was another flare-up of poaching in Mudumalai and Sigur, and this continued into the following year. Another elephant familiar to me fell to the poacher's guns—Cross Tusks (a different elephant to the one in the Biligirirangans), who had been made famous in Naresh Bedi's film 'Elephant: Lord of the Jungle'. Researchers Ajay Desai and Sivaganesan retrieved its skull. Cross Tusks was to them what Divergent Tusks was to Selvam and myself.

In April 1987, K. S. Neelakantan, a young forest officer, took over as Warden of Mudumalai and the poaching spree came to a halt.

Tragedy, however, struck at Satyamangalam. On 14 July 1987 Chidambaram and his men intercepted a truck carrying poached sandalwood. Chidambaram was shot dead at point-blank range, allegedly by Veerappan himself. When I read of his death in the newspapers I could not believe it. Chidambaram was tough and dedicated, and he had been a personal friend. Sadly the news was true. This was a tragic end to the life of a man dedicated to protecting our wild lands and its creatures. The following year, Chidambaram was posthumously awarded a gold medal by the President of India for his dedication to duty.

Even the shocking murder of Chidambaram did not sufficiently rouse the state governments to track down Veerappan. The poacher and his associates continued to roam the jungles at will and he had by now acquired the image of a Robin Hood among the villagers. Veerappan's intelligence network was far superior to that of the government and the co-operation of local villagers was essential if the authorities were to catch Veerappan. This was not forthcoming for two reasons. Anyone spilling the beans on the poacher's activities faced the real risk of being killed. Veerappan was also supposed to generously dole out money to his informants and helpers among the villagers.

That year I carried out a quick survey of the population structure of elephants in the Biligirirangans and the Nilgiris. When I tabulated the results I found that the proportion of adult males had indeed come down by 40 per cent since my previous assessment. This conformed almost precisely to what my population model had predicted—a reduction of the adult males from 6.5 per cent in 1982–3 to about 4.0 to 4.5 per cent of the total population by 1987. This validated my population model, but it was not good news for the elephants.

When I visited Hasanur I learnt from the farmers that Vinay had not troubled them during the previous two years. Either he had shifted his range of operations, or else he too had fallen to the poacher's gun. I suspected that the latter was the case, as Vinay was much too addicted to the *ragi* fields for his own good. He was a perfect example of what my good friend, Professor Lahiri-Choudhury, would almost affectionately call a *goonda*. Sadly, such *goondas* have had their day.

At Karapallam, the *Acacia suma* trees showed little sign of having suffered any damage during the previous few years. This clearly brought home the need for me to study ecological processes over a longer-term period and a larger spatial scale before coming to any definite conclusions about their dynamics.

This is precisely what I began to do during the following year. From 1988, the scene of my research shifted to the Mudumalai Sanctuary as part of the Nilgiri Biosphere Reserve programme. I began organizing a long-term programme of studying, among other things, the dynamics of elephants and trees in a tropical deciduous forest. Assisted by a team of researchers and funded by the Indian government's Ministry of Environment and Forests, I set about establishing a permanent plot of fifty hectares in which every plant above one centimetre stem diameter was identified, measured, tagged with a number, and its location mapped precisely. When the job was finally done a year later, we had enumerated nearly 26,000 individual living plants belonging to seventy-one species. Since then, the plot has been censused every year in order to study the changes in vegetation and the role of elephants and fire in the dynamics. During the first two years of the study the populations of two plants, *Kydia calycina* and *Helicteres isora*, have come down substantially, due in large measure to damage by elephants. The precise implications of this for the habitat are not clear at this stage.

The inspiration for setting up a plot on this scale came from

Steve Hubbell of the Smithsonian Tropical Research Institute and Princeton University. He had been the prime motivator behind a fifty hectare plot set up during 1980–2 on Barro Colorado Island in Panama. A plot on such a scale had never been attempted by ecologists before; one hectare was considered the standard by most researchers. When I went to Panama to see his plot, Steve argued convincingly why such large-scale plots were necessary in order to be able to study tropical forest dynamics effectively. I decided that we would repeat this scale of study in Mudumalai.

In April 1988, Sudha, Gitanjali and I packed our bags and moved from Bangalore to live at our field station at Masinagudi in Mudumalai. For me, it was exhilarating to be back full-time with the elephants and the forest, doing all that I really loved to do. The elephants were all there—tall ones, short ones, 'round ones, tusked elephants, tuskless elephants, timid ones and angry ones; the only type missing was the truly magnificent tusker. All the big tuskers I had earlier known were now gone, most of them mown down by the ivory poachers.

I quickly became acquainted with this new group of elephants. There was Blunt Tail, a tusker who promptly came to the Hambetta swamp every dry season, three adult *makhnas*, a large bull with only one right tusk, and even a junior replica of Vinay with a broken right tusk.

There were also three angry cows I called the Torone sisters (after Ian Douglas-Hamilton's four aggressive cows at Manyara, of which one had to be shot). The sisters were part of a sub-unit of five elephants, which was in turn part of a joint family of eighteen elephants.

These three cows were a highly emotional lot, trumpeting regularly and running hither and thither all the time. Sometimes only one or two of them would threaten, but at other times all three of them would charge together. The rest of the herd, including some older cows, were usually calm, although once the entire herd bunched together and advanced on me.

It is a truly unnerving and unforgettable experience to see eighteen elephants spread out flank to flank with ears extended, bearing down upon one. When this happened to me, my jeep, recalling earlier times, failed to respond. Fortunately, only one of the Torone sisters mustered the courage to advance all the way up to within striking distance. My companion botanist,

H. S. Suresh, looked on aghast, before retreating. I could not help bursting into laughter at this ludicrous event—ten years with elephants is no guarantee that you will not land yourself in a soup!

The work at Mudumalai also brought with it other responsibilities. I now had to guide a team of researchers and it was important to see that everything went smoothly. Our team ran into armed poachers several times. Surendra Varman, who was censusing large mammals, and his tracker Annadurai were attacked by a sloth bear, but fortunately the tracker escaped with only minor injuries and Surendra was unharmed. Botanist H. S. Dattaraja was also chased by a stampeding herd of elephants while working on the fifty hectare plot. Another time, R. Arumugam was sitting up a tree observing a dhole pack, when a cow elephant came along and decided to sample the shoots of parasitic fig that grew on a much smaller tree upon which his Kurumba tracker Shivaji sat. As the elephant's trunk came up to pluck the leaves it was practically within reach of the gasping Shivaji, who could climb up no further. Arumugam yelled at the top of his voice and fortunately the elephant decided to shift its attention to more down-to-earth items.

Living in Mudumalai, away from from the bustle of a city, was a completely new experience for my wife Sudha, but she quickly began to enjoy it. For our daughter Gitanjali, who was two years old when we shifted there, it was a wonderful and educational experience. From our bungalow at Masinagudi we could see spotted deer and occasionally elephants coming for a drink at the Maravakandy reservoir, which was close by. In the evenings we would go for a ride in the jungle or visit the elephant camp at Theppakadu and watch the animals there being bathed and fed. Gitanjali got well-used to seeing wild elephants and to taking rides on elephant back inside the jungle. She learnt to be silent and to talk only in whispers like the rest of us whenever we were watching animals and could recognize all the common species. The only problem was that she would insist on riding every elephant that she saw, whether captive or wild!

During the morning of 5 April 1989, while I was absorbed in some papers at my desk, I heard a commotion outside and shouts of '*yanai, yanai*'. Looking out, I saw Gitanjali running towards the elephants at the reservoir, along with a group of older children. I rushed out to catch her before turning my attention to the elephants. There were certainly no tame elephants at Masinagudi!

Sure enough, a herd of elephants had come to the farther side of the reservoir. We all went down to the water's edge and watched the herd for half an hour before returning to the house. Surendra Varman stayed back and continued to observe them. An hour later I sent our tracker Bomma to check with Surendra about when he would return. We had scheduled other work for that afternoon. Bomma soon returned and informed me that Surendra would come back after the elephants had finished 'crossing'. I assumed that the elephants had begun to swim across the waters of the reservoir. When Surendra finally returned he told me that a bull had arrived which later mated with one of the cows. He had asked Bomma to inform me about this. When I checked with Bomma why he had not told me of this, he insisted that he had indeed informed me that the elephants were 'crossing'. It was then that it dawned upon me that I had failed to grasp the euphemistic nuances of Kurumba English!

While observing elephants at Mudumalai, I began to frequently see males in the five to ten year age range on their own or in the company of other young males. It was not uncommon to see lone males even younger than five years. One juvenile hardly three years old was seen wandering alone for two days. Earlier, I had not seen such a phenomenon in the Biligirirangans or even at Mudumalai, an observation which was also confirmed by Selvam.

The reason why males are straying away from their families at such an early age is not very clear. One possibility is that the elimination of most of the large bulls by poachers has triggered this early dispersal. The young males can explore on their own without fearing the larger bulls. Another possibility is that the high density of elephants in Mudumalai is causing changes in the conventional social patterns due to increased competition; an aspect of elephant social organization which calls for further study.

During the few years of its existence, our field station in Mudumalai has been host to several students and scientists, including some very famous names—William D. Hamilton, John Bonner, George Schaller, Steve Hubbell, Egbert Leigh, and even a Nobel Laureate in Chemistry. Mudumalai's rich wildlife never disappointed anyone. When our institute's director, C.N.R. Rao, an internationally known chemist, visited us for a few days, several herds greeted him within a few minutes of his arrival.

In May 1989 the fifty hectare plot was nearing completion. Madhav came down from Bangalore to celebrate the achievement and we were also joined by P. Padmanaban, now the state's Chief Wildlife Warden, by Mudumalai's Warden K. S. Neelakantan and by Dr K. The dignitaries numbered the remaining few trees in the plot to complete the enumeration and later we held a party at Masinagudi in celebration. The next month we had to reluctantly return to Bangalore.

I was now increasingly involved in international efforts for conserving the Asian elephant. The previous year at a meeting of the IUCN's Asian Elephant Specialist Group in Chiangmai, Thailand, I had been chosen as the Deputy Chairman of the group. They had also recommended that a secretariat for the group be established at our department in Bangalore in order to co-ordinate its survey, research, and conservation efforts. The secretariat, called the Asian Elephant Conservation Centre, formally began operating in June 1989 with funding from World Wide Fund for Nature.

These new commitments and responsibilities meant that I was now busier than ever. The task was rather daunting. Considerable progress had been made in documenting the elephant populations of India and some other Asian countries, but virtually nothing was known about the elephant populations in countries such as Myanmar (Burma), Laos, Kampuchea, and Vietnam. We somehow had to make a breakthrough in these countries but, given the political conditions, it was going to be difficult. I was fortunate in having to work with our Chairman, Lyn de Alwis, who had retired as head of Sri Lanka's wildlife department, a person who was never at a loss for the right word for every occasion, and with our Executive Officer, the ebullient Charles Santiapillai, also from Sri Lanka, who had a very wide field experience in the Southeast Asian countries. Together, we set about organizing data bases, training workshops on elephant management, and field projects.

The next year, on a visit to the United States to work with Chris Wemmer of the National Zoo on captive elephant management, I had the opportunity to lecture at various places on Asian elephant conservation. While the African elephant had received enormous attention, the Asian elephant has been rather neglected by the international conservation community. This is rather ironic, as the Asian elephant has both enjoyed a much closer cultural association with people and its population numbers are far lower than that of the African species. Generally, most people either seem to have the impression that the Asian elephant is doing very well or that they have all been taken into captivity. I attempted to correct these impressions.

The elephant is a very powerful symbol in organizing conservation efforts in Asia. I realized this during a visit to Japan in August 1990 to attend the International Congress of Ecology. Most people I met there were very interested in elephants, even though the species has never lived there. I was one among a small group invited for an audience with Emperor Akhihito. The Emperor and his second son, Prince Akishino, both biologists, were extremely interested in the welfare of the elephant and fondly recalled memories of the elephant Indira which had been given to Japan by the former Indian Prime Minister Jawaharlal Nehru. I made an appeal that Japan should help other Asian countries in conserving their elephant populations.

By 1990 there were encouraging signs that the elephant populations of India were entering a brighter phase. From 1988 onwards there had been a noticeable decline in poaching of ivory in the southern states. This was, of course, partly because there were few male elephants remaining. The decline in poaching was also due to the increased vigilance of the forest department in some regions. The poachers have increasingly turned to other illegal activities, such as sandalwood felling, giving a respite to elephants. Veerappan, however, continued to defy the authorities. In spite of a massive hunt for him after he and his gang had ambushed and killed four of Karnataka's policemen in April 1990, he still eludes the net.

There have also been changes in the government's ivory trade policy. The high customs duty on the import of African ivory was abolished in 1988. By then, most of the dealers who had been importing legal ivory were frustrated with the difficulties in the trade and had turned to other products. Esmond Martin told me in early 1989 that the number of carvers in the country

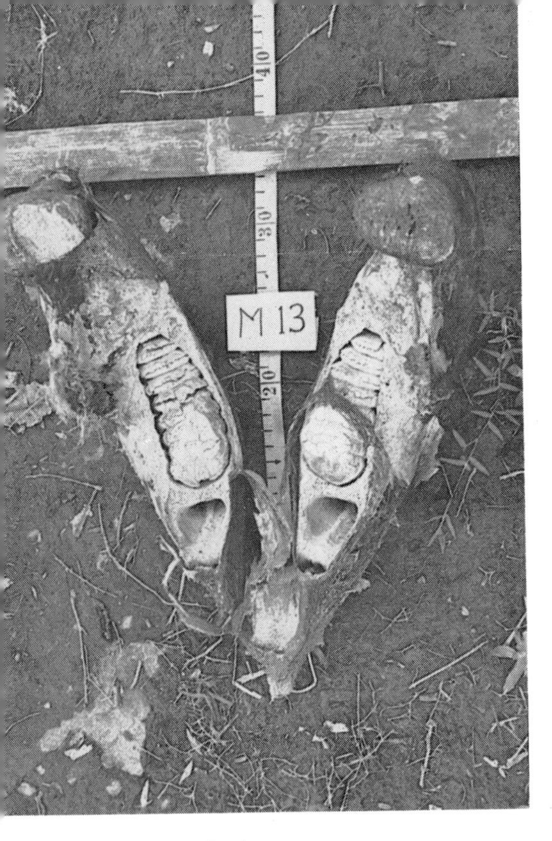

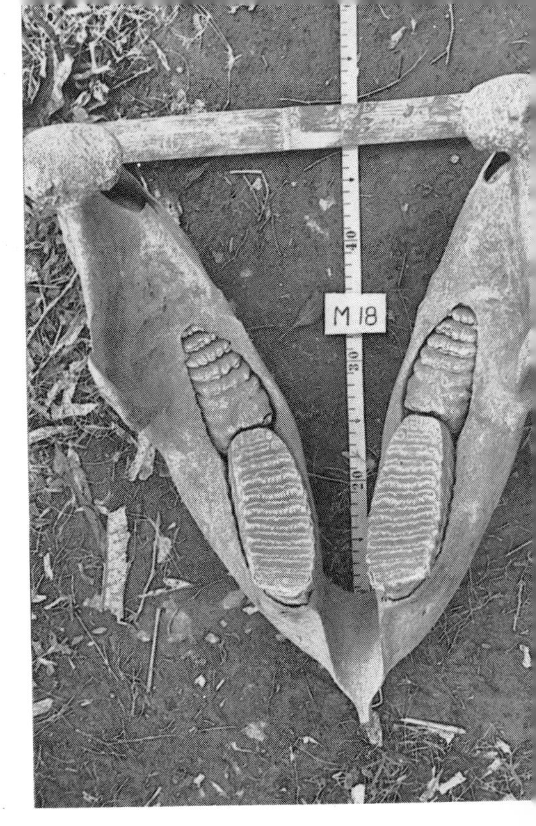

a) 1 year

b) 7 years

The lower mandibles of elephants aged approximately:

c) 15 years

d) 42 years

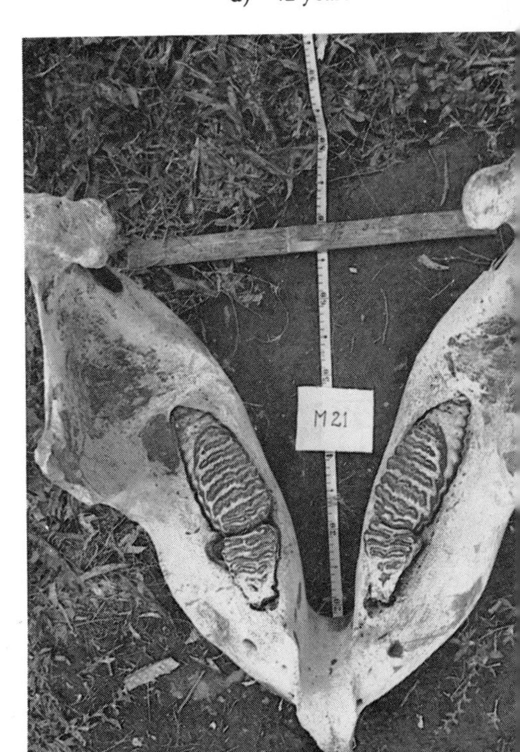

A young cow drinks from a pit she has dug on the Araikadavu during the drought.

Tara and her family during the drought of 1982.

The cow, striken with diarrhoea, struggles to stand up.

Tara, one of the oldest Asian elephants to have ever lived, photographed at Theppakadu in Mudumalai when she was seventy years old.

The Torone sisters and a part of their family advance flank to flank.

had declined to only 2000, though this is still the largest number in the world, just above that of China. The historic ban on African ivory imposed at the October 1989 meeting of CITES (the Convention on International Trade in Endangered Species of Flora and Fauna) also dealt a further blow to the carving profession. This was in one sense a pity because an ancient tradition of art was being eroded. The best option would have been to maintain the carving profession through the supply of legal ivory from elephants which had died naturally, but this option had been swept aside by the tidal wave of poaching.

In 1990, the Indian government appointed a seven-member task force for Project Elephant under the Chairmanship of S. Deb Roy, a veteran forester from Assam, who was now the head of wildlife in Delhi. Along with two colleagues, J. C. Daniel and D. K. Lahiri-Choudhury, of the Asian Elephant Specialist Group, I served on this task force. This initiative was especially heartening; it indicated a strong commitment by the Indian government to preserve the very soul of the country's rich cultural and biological heritage.

The future of the elephant population now depends on whether or not the habitat can be preserved and whether the lull in poaching is sustained. If the habitat continues to shrink, the elephants will have nowhere to go. If tuskers continue to be killed at the present rate there is a danger that there will be too few of them left to breed, resulting in a decline in fertility. The tuskless males will, of course, be at an advantage and will contribute their genes in greater proportion than will the tuskers to future generations. If such a scenario were to unfold, the *makhnas* will come to gradually dominate the population. This might take several elephant generations, given the slow rate at which elephants multiply and die. The elephant itself might continue to survive as a species . . . but then, to many, the beauty of an elephant is in its tusks.

CHAPTER 9

The March of Time

> On the border of the kingdom, the king should establish a forest for elephants guarded by foresters. They should kill anyone slaying an elephant.
>
> The time for catching elephants is summer. A twenty year old should be caught. A calf, an elephant with small tusks, a tuskless male, a diseased elephant, a female elephant with young and a suckling female elephant should not be caught.
>
> Kautilya's *Arthasastra* (c. 300 BC to AD 300)

The calf was extremely frightened by now. For the past three days and nights she had been wandering in search of her family, feeling lonely and miserable. Dark clouds still hung menacingly in the sky, but at least it was now not raining. Earlier, the very heavens had opened to deluge the earth in a show of fury, orchestrated by lightning and thunder. Her problems had begun when her family had decided to cross a swollen river. It had been a bad miscalculation on her mother's part. The rapidly flowing waters had swept her away, while the others stood by, helpless. She had been fortunate to survive, having been swept into a relatively calm pool from which she had managed to swim to safety.

It was quickly becoming dark when she smelt a familiar smell. She was frightened and stayed rooted to the spot. Within a few moments she saw familiar shapes moving towards her and hopefully took a step forwards. Cattle. The cattle stopped in their tracks and eyed her rather suspiciously; after all they had in the past often been chased away from ponds while having a drink, by elephants trumpeting shrilly and rushing at them threateningly. The cattle moved on quickly, not knowing that a new companion was trailing them. It was only when they entered their pen, an hour later, that they noticed the elephant calf in their midst. They reluctantly let her stay for the night—at least this one did not seem to mean them any harm.

The next morning dawned bright and clear. There was great

excitement in the village. All the people flocked to see the latest arrival. Some of them were familiar with elephants. They had encountered them in the eastern forests, but had usually avoided the huge beasts. Here was a small, frightened but friendly elephant holding out its trunk, smelling each of them in turn. The bolder lads clambered onto the elephant's back as they always did with their cattle. Over the following weeks the elephant calf became everyone's darling. She accompanied the cattle for grazing, she gave rides to the village children, she even carried back small bundles of wood the grazier took home in the evenings. The village may have been one of the hundreds of settlements dotting the vast Indus river basin. The year—around 2000 BC.

This marked the first step in a unique relationship between beast and man, an interaction that has remained unsurpassed for its sheer contrast and splendour, and is unlikely to be overshadowed by any association that may develop in the future. The elephant has both been worshipped as an incarnation of god and brutally slaughtered for its meat and tusks; it has trampled people and their crops but has also carried them and their heaviest burdens; kings have craved to keep them in their armies and have also offered them as ambassadors of peace.

The seals of the ancient Indus civilization depict many animals, among them the elephant. The elephant is shown with a line extending from the top of its back down to just behind its hind leg. This has been interpreted in the words of Richard Carrington as 'the forward edge of some kind of saddle or drapery', a fore-runner of the more elaborate *howdah*.

D. K. Lahiri-Choudhury, an authority on the elephant in ancient literature, is rather sceptical about this interpretation. Pulling out an enlarged picture of a seal he pointed out to me that the line was curiously flush with the lower portion of the elephant's body, almost as if it had been carefully tailored. If this indeed was drapery, then it should have been hanging loose. This was not to say that captive elephants were not present during the Indus civilization. There were many Indus seals with figures of elephants and some do have ropes on them, though none has a proper harness or a rider. He concurred that the elephant was first tamed by the pre-Aryan people of the subcontinent.

One of the earliest references to domesticated elephants is in the *Ramayana*, the great Indian epic, where the description of the forces arrayed against Rama, presumably those of Dravidian people, mentions the use of elephants.

If the Dravidian people captured elephants, they seem to have done so by trapping them individually or in small numbers in pits, a method that prevailed in the south of the country until the recent ban on their capture. Around 1500 BC the so-called Aryan people from West Asia entered the subcontinent. By this time the Indus culture had already lost much of its earlier glory. The Aryans realized that the great beasts would be invaluable in their conquest of the northern Indo-Gangetic plains. They captured elephants in even larger numbers through *kheddah* or driving entire herds into stockades. Once tamed, the captive elephant was turned against its wild relatives. Trained elephants are essential for capturing wild elephants and were used for hauling timber to establish settlements all over the vast Gangetic river basin.

Lahiri-Choudhury also points out that ancient literature, such as the *Rig Veda* (1500–1000 BC) and the *Upanishads* (900–500 BC) associated with the Aryans, contain many references to elephants. One remarkable manuscript, the *Gajasastra* (or elephant lore), is traced to sage Palakapya who lived in Bihar during the sixth to fifth centuries BC. The story goes that Gunavati, a demi-goddess, was changed into an elephant by the curse of the sage Matanga, and she gave birth to Palakapya, who from birth lived and wandered in the jungles among wild elephants. When the kingdom of Anga was troubled by crop-raiding elephants, its king Romapada had them captured and chained up with the help of many sages. A grief-stricken Palakapya came and pleaded with the king to extend his protection to elephants. His narration to the king about the natural history of elephants is the basis of the *Gajasastra*.

Another great Indian epic, the *Mahabharatha* (c. 1000 BC), contains one of the earliest known references to the use of elephants in war. In their battle with the Kauravas for Hastinapur (literally elephant town), the Pandavas have to find a way to eliminate Dronacharya, the veteran warrior-teacher of both the clans, who was arrayed on the side of the Kauravas. Krishna, the master strategist of the Pandavas, suggests that Drona be told that his son Aswathamma is dead. Drona at first refuses to believe this unless it is confirmed by Yudhishtira, the eldest of the Pandava brothers, known for his steadfast adherence to truth. Yudhishtira protests that he cannot utter a lie, whereupon his younger brother Bhima picks up his mace and clubs an elephant named Ashwathamma to death. Yudhishtira now pronounces to Drona, 'Ashwathamma is dead, Ashwathamma the elephant'. The last word, feebly uttered, is drowned by the sound of Krishna beating his drum. In any case, Drona had already collapsed upon hearing that his son was dead.

It is doubtful whether the elephant was ever a decisive factor in the outcome of a major battle. Alexander exposed the vulnerability of an elephant army to horse-based cavalry at a place not far from where the wild elephant may have been first tamed. When he reached the banks of the Jhelum river, a tributary of the Indus, in 323 BC, he was confronted by an impressively arrayed 200-elephant army of King Porus, ruler of the Punjab. The weary Macedonians were at first terrified at the sight of the elephants, but Alexander decided to press on with a brilliant strategy. He crossed the swollen Jhelum upstream at night and encircled the elephant army. The Macedonians archers picked off the mahouts, while other men hacked away with spears and scythes at the elephants in order to create confusion in their ranks. The enraged elephants became uncontrollable, trampling both friend and foe alike. The Macedonian horsemen finally carried the day and even King Porus, seated on a huge elephant, fell down unconscious. In spite of his victory, Alexander seems to have been impressed by the splendour of his opponents, for he allowed Porus to retain his kingdom. It is said that the Macedonians decided that they had had enough in battling against the elephants and retreated once they reached the river Beas. Greek historians later made up stories of fierce elephants held by other rulers in order to justify this mutiny.

Yet elephants continued to be the prized possessions of kings for another 2000 years. The Mauryan kingdom maintained a large elephant army, even importing elephants from distant lands of the subcontinent. Chandragupta Maurya, who repulsed Alexander's successor Selucus Nicator, had 9000 elephants in his army. The *Arthasastra*, a manual of statecraft written by Kautilya around 300 BC during the Mauryan period (though additions are believed to have been made until AD 300), gives instructions on the capture and maintenance of elephants. While it urges the elimination of elephants from river valleys which are to be brought under settlement, it also advises the setting up of sanctuaries for elephants in the hill forests along the periphery of the kingdom. Anyone killing an elephant in these sanctuaries was to be put to death, presumably because they provided elephants for the king's armies.

This dependence on the elephant for transport and for their armies by the rulers of the more advanced civilizations which cultivated the river valleys, may have led them to impose a taboo on the killing of elephants for meat, which was probably the practice of the hunter-gatherers and shifting cultivators inhabiting the hill forests. Elephants were more useful alive than dead. Prior to this, the more primitive societies did consume elephant meat; the ancient Tamil Sangam literature (first to fifth centuries AD) refers to this habit among the hill tribes in the south. This can still be seen in two tribes in the northeast; the Nagas and the Mizos, who did not have much to do with more modern societies until relatively recent times.

By the dawn of the Christian era, the elephant had disappeared from a substantial portion of the Indo-Gangetic plains and possibly from the southeast of the subcontinent. The *Arthasastra* describes eight *vanas*, or forests, as the abodes of elephants. Historian Thomas Trautmann places these in northern India, eastern India, central India and western India. The southwestern part of peninsular India has been strangely omitted, but this could simply be because the Mauryans and latter-day compilers of the *Arthasastra* were not familiar with this region. The Tamil Sangam literature of this period certainly mentions the abundance of elephants in the land of the Cheras, or present-day Tamil Nadu and Kerala.

During the third or fourth centuries there arose a powerful new symbol for the veneration of elephants—the image of Ganesha, the remover of all obstacles, the elephant-headed god of the Hindu pantheon. The mythology regarding the origin of Ganesha is rather confused. In one popular version, the goddess Parvathi formed a *gana*, or attendant, out of clay and adopted him as her son. One day the *gana*, who was guarding the apartments while Parvathi was bathing, refused entry to her consort Shiva, one of the supreme gods in the Hindu triumverate. An enraged Shiva chopped off the *gana's* head, but later, so as to avoid Parvathi's wrath, he sent an envoy in search of a replacement. The envoy returned with the head of a one-tusked elephant which Shiva used to restore the life of the *gana*. For the latter's devotion to duty (and possibly because Shiva needed to make peace with Parvathi!) he was made leader of Shiva's *ganas*.

According to another version, Shiva himself told his son that while they were taking a walk through the jungle, he and Parvathi had seen two elephants mating and had been excited by this and encouraged to do the same. Thus their son was born with an elephant head.

A third story attributes Ganesha's broken right tusk to his fight with Parasurama, an incarnation of Vishnu, who also ranks amongst the three supreme Hindu gods. A fourth version gives credit to Ganesha for writing the *Mahabharatha* non-stop, at the dictation of the sage Vyasa, by breaking his tusk and using it as a pen.

An elephant-headed god is not peculiar to the Hindus. In fact, myriad forms of Ganesha are found in countries as far apart as China, Kampuchea and Japan. The Buddha himself is considered an incarnation of a sacred white elephant. When Queen Maya lay dreaming in her sleep, a white elephant entered her womb and she later gave birth to Siddhartha, the Buddha.

The current image of Ganesha as the 'remover of obstacles' is actually a complete transformation which evolved over eight centuries from his earlier image, under the epithet of Vighnaraja or Vinayaka, as the 'creator of obstacles'! From the fifth century BC until the third century AD, Vinayaka is actually depicted in mythology as a malevolent spirit, part of a pantheon needing to be pacified in order to avoid personal ills. As historian of religion G. S. Ghurye states, 'this last problem [i.e. transformation] defies a perfectly rational and reasonable explanation'. I think the explanation may be quite simple, however. In earlier times, the wild elephant would have been a threat to human life and to crops over most of the subcontinent; the *Gajasastra*, for instance, describes the ravages of elephants in the kingdom of Anga. With the gradual elimination of wild elephants over the Indo-Gangetic basin by the Aryan settlers, and the growing significance of elephants in armies or simply as beasts of burden, there was no further need to consider elephants as evil spirits; in fact, the very opposite would now have been true. Vinayaka, the malevolent elephant-headed spirit and creator of obstacles was, therefore, transformed into Ganesha, the benevolent elephant-headed deity who had the power to remove all obstacles. It is interesting to note that the worship of the benevolent Ganesha arose among the élite, who were the least likely to suffer from elephant depredation, and was only grudgingly accepted by the common people, many of whom, at the interface of jungle and cultivation, would have faced a threat from the elephants.

The worship of Ganesha reached the south of the country during the seventh century AD when the Pallava dynasty held sway over the region. With this, the killing of elephants for meat would have been given up over practically all the Indian subcontinent, save the hill forests east of the Brahmaputra river valley. Although people did not eat elephants, they certainly continued to capture them and to push them out of river valleys they colonized.

The trade in ivory and elephants is a complex affair. As early as the sixth century BC, African ivory from Ethiopia was being sent to India according to historian E. H. Warmington. The volume may not have been large initially but certainly was substantial after the dawn of the Christian era. Carved ivory from India was exported to the West, including to Greece and Rome. At one stage, the European trade in Indian ivory seems to have been even larger than that in African ivory. The Indian rulers also imported elephants from the island of Ceylon and from Pegu in Burma.

By the end of the tenth century, the elephant had disappeared from most of the northern Indo-Gangetic plains, the river valleys in the southern peninsula, and the coastal tract.

The passion of the Hindu kings for elephants seems to have passed on to the Muslim rulers who gained ascendancy over the northern part of the subcontinent from the eleventh century onwards. In AD 1031 the Ghaznavid kingdom possessed 1670 war elephants. The Sultans of Delhi who ruled from AD 1192 to AD 1398 also considered it prestigious to have a large *pil-khana*, or elephant stable. At the peak of their power they had about 3000 elephants, of which some 1000 may have been fit for battle. When Amir Timur entered Delhi in AD 1398 to overthrow the Malikzada Sultan Mahmud the *pil-khana* had been reduced to just 120 war elephants.

The introduction of gunpowder into warfare during the fourteenth century and in artillery and hand-guns during the sixteenth century almost completely eroded the direct utility of elephants in battle. Surprisingly, the Moghul rulers of the sixteenth and seventeenth centuries continued to acquire elephants in large numbers. During this time there still existed considerable numbers of wild elephants in central India, as far west as Dohad in present-day Maharashtra, which the Moghuls depleted through capture. The Moghuls did not seem to kill elephants in their sport hunts; their elephant hunts were probably confined to capturing them. Jehangir (AD 1605–27) is reputed to have maintained a stock of 12,000 elephants in his army, a level which implies the over-exploitation of wild stocks. The elephant certainly disappeared over a large portion of central India during the Moghul period.

The distribution of wild elephants at the end of the Moghul rule seems to have remained largely unchanged until the middle of the nineteenth century. One population extended along the Himalayan foothills into the hills of the northeast, another large population roamed over the Western Ghats and certain adjoining hills of the Eastern Ghats in peninsular India, while a third smaller population was confined to the eastern portion, primarily Orissa and Bihar, in central India. During this period the British penetrated the hill forests and began cultivating tea and coffee on a large scale in these hills. They also contributed their share to the decline of wild elephants through their sport of hunting 'big game'. The killing of elephants for sport had not been a part of the Indian ethos for several centuries prior to this. One British planter is reputed to have shot about 300 elephants, most of them cows and calves, in the Wyanad of southern India.

The Elephant Preservation Act introduced in 1873 in the Madras Presidency brought some relief to elephants in a portion of the British territory in India, but elsewhere they continued to suffer persecution. Some of the Indian rulers of the princely states, in their eagerness to be considered as equals to the colonial rulers, also began hunting elephants and other animals for sport. The capture of wild elephants continued unabated; the figures I have compiled suggest that 30,000 to 50,000 elephants were captured or shot in the subcontinent, largely in the northeast, during the period 1868 to 1980. This figure could easily be as high as 100,000 for Asia as a whole during this period.

India's independence in 1947 gave a new fillip to industrialization. Dams have submerged river valleys in the hill forests, mines have stripped entire hill slopes bare, and the burgeoning human population has pushed further into the forests in its hunger for producing more food. The colonization of the terai or moist forests along the Himalayan foothills has separated the northwestern and the northeastern elephant populations. The country's forests have shrunk by thirty per cent since independence. However, realizing the plight of the elephant, the Indian government has now launched Project Elephant to save this magnificent species from extinction.

Over the past decade and a half, the members of the Asian Elephant Specialist Group of the IUCN, assisted by the WWF, have assembled the approximate numbers to give a picture of the elephant's distribution. This has not been at all easy. In some countries the political situation has made it impossible for anyone to carry out field surveys. Even where field-work has been possible, the estimation of elephant numbers in dense, equatorial rain forest is usually merely an 'educated guess'.

Robert Olivier was the first person to piece together details of the elephant's distribution and status throughout Asia in 1978. Since then, we have an improved picture of many regions. The elephant is largely confined to the hill forests of the Indian subcontinent, continental Southeast Asia, and certain islands of Asia.

India is believed to hold the largest number of wild elephants, estimated at 18,000 to 24,000. The forests of the Shiwaliks and the terai along the Himalayan foothills in Uttar Pradesh hold an isolated population of about 750 elephants. Some of these also move into adjacent Nepal, which has very few resident elephants. In the northeast elephants are distributed more extensively along the sub-Himalayan tract extending from West Bengal through Bhutan, Arunachal Pradesh and Assam. While 3800 to 5800 elephants may be found there, other populations include one of 2500 to 3500 in the Garo-Khasi hills of Meghalaya and about 1900 elephants in the floodplains of the Brahmaputra, chiefly in Kaziranga, and the Naga hills. The extensive shifting cultivation practised in the hill forests here is a major threat to elephants. The period of rotation of sites has shrunk over the years from over thirty years to less than five years, arresting the regeneration of forest and keeping the hill slopes permanently devoid of any tree cover. The considerable loss of habitat has escalated the conflict between elephants and people. Since major portions of the forests are owned by village communities, these are not subject to government regulation or management.

In the central Indian states of Bihar and Orissa there are a number of small populations totalling 1500 to 2000 individuals. Most of these are in degraded, fragmented habitats and they do not seem to have a secure future. The estimated 8,000 to 10,000 elephants in the southern India range are seen as nine or more distinct populations over the Western Ghats and the adjoining hills of the Eastern Ghats in the states of Karnataka, Kerala and Tamil Nadu. Their habitat is mostly government-owned reserve forest, although numerous hydro-electric and irrigation dams, monoculture plantations, mining, and pockets of cultivation have made considerable inroads. Ivory poaching has also been a serious problem here.

Bangladesh has less than 250 resident elephants in the Chittagong hills, with another 100 or so elephants shared with Myanmar and India.

The last of the elephant herds in China are found in the south of the Yunnan Province bordering Myanmar and Laos.

Myanmar (Burma) may still hold some surprises. Current estimates of the elephant population range between 3000 and 10,000 animals, but these estimates do not have any objective basis. Elephants are still widely distributed, the major areas being Myitkinia-Upper Chindwin to the north, Arakan Yoma to the west, Pegu Yoma in central Myanmar, and Tenasserim in the south of the country. No authentic contemporary information on the conservation of elephants in Myanmar is available; this is likely to remain the situation for some years to come given the prevailing political problems there. Recently, there have been disturbing reports of the Karen people in the northeast poaching elephants and selling their products in neighbouring Thailand.

Once famous for its elephant traditions, such as that of the sacred white elephant, Thailand has lost much of this past glory. Hardly 3000 to 4500 elephants were believed to survive in the country in 1977. Today the situation is even more bleak. Many of the elephants here are found in the Tenasserim range along the border with Burma. Elephants are also seen in the Petchabun mountains to the northeast, the Dangrek mountains bordering Laos and Kampuchea, and in peninsular Thailand to the south. The magnitude of the decline can be gauged from the fact that only seven protected areas are now thought to have populations above 100 elephants.

Elephants in Laos, Kampuchea and Vietnam are largely

found along the international borders between these countries. In addition, the Cardamon and Elephant mountains in the southwest of Kampuchea are also potential habitat. Information from Laos and Vietnam is beginning to trickle in, but Kampuchea is still an unknown situation. No one has evaluated the impact of the wars on elephants in this region. The available figures suggest a total of 5500 to 7000 wild elephants for these countries.

Most of the elephants in peninsular Malaysia are scattered in small herds or even as solitary bulls. Only the Taman Negara and Endau Rompin regions seem to have viable populations. The cause of the elephant's decline here is the massive expansion of commercial agriculture, chiefly oil palm and rubber plantations, which are replacing the natural rain forests. Mohd. Khan bin Momin Khan, the redoutable head of peninsular Malaysia's wildlife department, who has done much to save the remaining elephants, estimates that only 1000 elephants survive in peninsular Malaysia.

On the island of Borneo, the elephant is found to the northeast in Sabah (a state of the Federation of Malaysia) and a small contiguous area of Kalimantan (Indonesia). It is still a mystery whether the estimated 500 to 2000 elephants here are indigenous or are descendants of captive elephants presented to the Sultan of Sulu during the mid-eighteenth century. As in peninsular Malaysia, the rapid clearing of natural forest for commercial agriculture is the main threat to their survival.

The story is similar on the island of Sumatra (Indonesia). The rain forests here are being felled for resettling people from the crowded island of Java under a transmigration scheme, and for raising plantations of oil palm, rubber and sugar cane. Charles Santiapillai and Raleigh Blouch who have studied the elephant situation here estimate that 2800 to 4800 elephants are scattered in forty-four different populations, with only five populations having more than 200 elephants.

Descendants of captive elephants roam over Interview Island, part of the Andamans and Nicobars (India) in the Bay of Bengal. These elephants would be of great interest to a geneticist because they are a mixture from different regions of Asia, including Assam and Myanmar.

The elephant lore of Sri Lanka is almost as ancient as that of India. However, the great herds were decimated through slaughter in the name of sport during the nineteenth century. Lyn de Alwis and A. B. Fernando, who have done much to save Sri Lanka's elephants, estimate that only about 3000 elephants currently survive in the dry jungles to the north and east of the island country. As elsewhere in Asia, the imperatives of development are fragmenting the habitat and isolating the elephant herds. Under the massive Mahaweli Ganga Development Project a number of large dams are being constructed and new land opened up for agriculture in the country's largest river basin. Although a network of protected areas has been provided for under the project, the elephant will still lose considerable ground. The current continuing political strife has made it difficult to reassess the elephant's status.

The figures compiled add up to a minimum of about 37,000 and a maximum of 57,000 wild elephants in Asia. Even considering that 15,000 elephants survive in captivity, the total population of the Asian elephant is just 10 per cent of the estimated numbers of the African elephant. The reasons for the decline of the two species have been historically quite different—the African elephant has been periodically decimated by ivory hunters, while the Asian elephant has been captured for domestication or eliminated over vast areas through the expansion of human settlements.

What does the future hold for the Asian elephant? The future may be largely tied up to the so-called 'protected area' system in the various countries. Only thirty per cent of the estimated half a million square kilometres of the elephant's range is part of the conservation areas of sanctuaries and national parks, and the elephant is certainly going to lose habitat outside these conservation areas in many regions. D. K. Lahiri-Choudhury estimates that only about fifty per cent of the present habitat for elephants will survive in northeastern India, given the inevitable loss to shifting cultivation. Some elephants will, of course, survive in the remote hill forests which are thinly populated by hunter-gatherers or shifting cultivators, which are currently outside the conservation areas. On the other hand, many conservation areas are not adequately protected and are merely reserves on paper.

Planning for elephant conservation has to consider two time scales. Our immediate goal is of course to ensure that the Asian elephant does not become extinct in the wild within the next fifty or hundred years; the ultimate goal should be to maintain its full evolutionary potential.

In the short-term, a small elephant population may become extinct due to chance events alone. A catastrophic event such as a drought or a disease epidemic could cause the death of all individuals in the population. Even under normal circumstances a small population might suffer from sheer 'bad luck'—all the individuals might die within a short span of time, or all could happen to be of the same sex! The probability of this happening becomes progressively smaller the larger the population size.

There is also a lot of inbreeding or mating between closely related individuals in a small population. We know from the experience of animal breeders that too much inbreeding results in genetic abnormalities, reduced growth, or even the sterility of the offspring. Small populations retain only a certain fraction of the genetic variation prevalent in the larger populations from which they were derived; even this variation is quickly lost through a chance process known as 'genetic drift'. The consequences of inbreeding or loss of genetic variation vary from one species to another; for elephants we have no precise information.

Apart from ensuring that elephants live through the next century, the larger goal should be to see that elephants do not reach an evolutionary dead-end. The story of life on earth is the story of evolution—of the fantastic proliferation of millions of species, of extinctions, but also of the continual adaptation of creatures to changing environments and their eventual transformation to entirely new species. Natural selection is the force that drives evolution, and selection operates on variation within a population. There is still controversy about the exact link between the evolutionary process and the extent of genetic variation in a population, but the complete loss of genetic variation may mean an end to the evolutionary process through natural selection. New variation can arise through genetic mutation but this is believed to occur at a very slow rate.

The spectacular evolutionary history of the Proboscidea (the order to which elephants and elephant-like mammals belong) beginning with *Moeritherium*, a small swamp-bound creature known from fossil finds in Egypt, which lived during the Eocene, about sixty million years ago, has left only two living representatives, the Asian elephant and the African elephant. Two close relatives, the mammoth and the mastodont, are believed to have become extinct in part due to 'overkill' by Pleistocene man. It would be a sad reflection on man's stewardship of the earth if the two living elephants are allowed to meet a similar fate.

Uncertainties concerning our scientific understanding of the evolutionary–extinction process should not be reason for inaction. It is wiser to err on the safe side when it comes to details about the minimum viable population size or the area of a reserve for elephant conservation. In the years to come we will increasingly have to deal with small isolated populations of the Asian elephant. Wherever feasible, conservation areas should allow for the migration of elephants between populations through connecting 'corridors'. If corridors are not possible, we may have to resort to capturing and translocating some animals between populations each generation in order to maintain their genetic variation. Such experiments may be tried initially in a few of the numerous small populations we have today. As our understanding of the links between genetic variation, evolution, and extinction increases, newer strategies for the conservation of small populations will no doubt emerge.

We need to do a lot more research on the genetics of elephant populations and for this purpose the large number of captive elephants can surely help.

At present there are less than a dozen Asian elephant populations known to have over 1000 individuals each and a sufficiently large area of habitat to support them. These populations include those in the Nilgiri–Eastern Ghats and Anamalai–Periyar in southern India, Arunachal–Assam and Kaziranga–Naga hills in northeastern India, southeastern Sri Lanka, Myitkina–Bhamo and Irrawady–Chindwin in Myanmar, Tenasserim mountains along the Myanmar–Thailand border, Laos–Kampuchea–Vietnam borders, and possibly Sabah in Borneo. In addition, some of the smaller populations, including that of Taman Negara in peninsular Malaysia and those in Sumatra are important as they may represent genetically distinct races that have evolved in the rain forests. Effort towards the long-term conservation of the Asian elephant should obviously pay more attention to these populations, and the further fragmentation and degradation of habitat should be prevented.

The range of the Asian elephant covers a wide variety of vegetation types, from equatorial rain forest in Borneo, Sumatra and peninsular Malaysia to semi-arid scrub in parts of Sri Lanka and India. It is essential that a conservation strategy spans this entire range. This would ensure the survival of the distinctive geographical races of the Asian elephant that have evolved over this wide area.

Can elephants and people co-exist peacefully? Can the need for natural resources be met without detriment to the elephant populations? I believe that a certain level of human exploitation of forests for timber, fuel wood or fodder is not incompatible with either the conservation of elephants or the overall biological diversity. The selective logging of moist forests does not cause any adverse vegetational change for elephants—in fact, it might even create a more favourable niche for large herbivores. Elephants are known to prefer secondary forest to primary moist forest. A vast proportion of the elephant's present range consists of forests that have been exploited to some degree in the course of human history, and elephants continue to do well in such secondary forests. It is the complete loss of forest habitat which decimates elephant populations. It is human greed that has led to the present environmental crisis. The recent ban on logging in Thailand, in response to devastating floods in the country, is a pointer to other tropical countries which are destroying their precious natural heritage.

We should also take sufficient steps to minimize the impact of elephants on people. In developing countries poor farmers bear the brunt of depredation. Their resentment towards conservation is understandable. They are more likely to accept the creation of a reserve for the protection of elephants if provided with relief from the depredation to their crops.

It is certainly not an easy task to keep elephants away from crop fields, however. Trenching along the forest–cultivation boundary has been tried in some regions, but this has many disadvantages. Trenches are expensive to dig and maintain; they may fill up during the rains if the soil is loose. Elephants may also fill up a trench by digging up the soil with their front feet in order to cross over it.

In recent years, a high-voltage electric fence, which delivers a severe jolt but does not cause any actual harm to animals or people, has become popular. This type of fencing has apparently achieved eighty per cent success in keeping out elephants from oil palm and rubber plantations in Malaysia. This fence design is also being tried in India, but improper construction and maintenance have so far contributed to its failure. Elephants cannot always be bluffed by such tactics either. Male elephants learn that their tusks, which are non-conductors, can be used to break the fence wire! Even tuskless males have a solution. One *makhna* at Masinagudi used to rear up on its hind legs and press down the wire with the soles of its front feet (which are poor conductors of electricity). Elephants even push over trees onto fences that obstruct their passage to the juicy sugar-cane field! The elephant's ingenuity has so far proved to be more than a match for man's technological gadgets.

Various other suggestions have been made as to how to keep elephants away from cultivated land. The use of specific sound frequencies or chemical repellents would be useful. Alternatively, the use of sounds or chemicals to lure elephants away from the crop fields may serve the same purpose. Perhaps the infrasonic language of migration or danger could be used to attract herds away from fields into the jungle. Chemicals or sounds that mimic the inviting signal of an oestrus cow could be used to draw bulls. However, I shudder at the thought of how the bulls might redirect their aggression if they consistently find no receptive cow waiting for them!

There is no choice, however unpleasant this may be, but to cull elephants that consistently raid crops or kill people. Unless there is no other possible solution, family herds should not be culled because such culling can easily result in negative population growth rates. The family herds, comprising some thirty elephants, that have dispersed into new habitat in the state of Andhra Pradesh in southern India and the herd of sixty to seventy elephants in Bihar in northern India, which have both been creating havoc in places far from their original homes, are obvious candidates for removal, however. Such herds are unable to happily coexist with people unless confined within expensive elephant-proof barriers, as has been done with the elephants of Addo in South Africa. Though it may be possible to capture some of these herds, with the larger groups this is an almost impossible task.

On the other hand, I believe that many elephant populations can be managed so as to reduce conflict with people by selectively culling particularly notorious male elephants. This would not have any detrimental impact on the population growth. My work clearly shows that, as expected from basic evolutionary theory, the average male elephant comes into far greater conflict with people than does an average member of a family herd. Considering that an adult bull elephant raids crops six times more frequently than does the average herd member, consumes twice as much crops per raid by virtue of its larger body size, and also damages the more economically valuable crops, the monetary value of damage by a bull is twenty times greater than that of the average herd member. Other than in exceptional cases, such as those already mentioned in Andhra Pradesh or Bihar, where herds indulge in manslaughter, the bulls are responsible for the overwhelming majority of killings. The removal of a bull would thus be more effective than culling larger numbers of a family group in reducing conflict. Further, in the elephant's polygynous society, the loss of a certain proportion of the males does not normally affect the fertility of the population. The remaining males ensure that all the oestrous females conceive. A population's growth rate depends essentially on the reproductive life span, the calving interval, and the death rates of the female segment of the population. However, if there are too few males, the fertility will of course decline. Here again, my studies show that an adult sex ratio of one male for every five females ensures normal fertility in the population.

Since most Asian elephant populations are not affected by the selective poaching of male elephants for ivory to the extent my study population was, the sex ratio will usually be less unequal. Populations in Sri Lanka and northeastern India, having large proportions of tuskless males, could certainly tolerate some removal of surplus males. However, it would not be wise to cull adult bulls in populations such as that of southern India, where poaching has already depleted their numbers.

There is one further problem which has to be considered in culling male elephants. Fewer males mean lower genetic variation. If too few males are available for breeding a situation akin to inbreeding will arise in future generations, and the undesirable effects of an unequal sex ratio are more likely to be felt in small populations than in larger ones.

When I wrote an article on this theme and sent it to an international journal dealing with conservation issues, one of the anonymous referees dismissed my arguments as merely of theoretical interest and 'moot'. The referee took the stand that we should totally keep our hands off any endangered species. He was no doubt sincere, but to me this seemed a perfect example of living in an ivory tower (I cannot help the pun!). Later, the paper was accepted for publication in *Biological Conservation*.

The recommendations in the *Arthasastra* on capturing elephants virtually amounted to selectively capturing only adult male tuskers that had attained maturity!

I personally lean heavily on the side of the non-interventionists when it comes to managing our wildlife. But the elephant makes such a serious impact on the lives of so many people that it would be foolhardy to close our eyes to reality. I am not recommending a free license for culling, but believe that the judicious culling of notorious elephants will, in fact, promote the conservation of the species by making it more acceptable to people. Vinay, the Akkurjorai Bull and their likes have to go, whatever one's own emotional attachment to them. Otherwise, people will continue to act on their own behalf, electrocuting elephants, shooting them down or, worse still, injuring them and then allowing them to live on in pain. Best of all, with Asian elephants one usually has the attractive and socially acceptable option of capturing them alive rather than having to kill them, as is being done in Africa.

We live in a world full of contradictions, where there is an obscene disparity between rich and poor, and where there are tugs and pulls in every direction, where there is a need for the have-nots to catch up with the haves, and a need for both modernization and conservation. As we recklessly continue to assault the earth's living systems, our hearts are sometimes moved by the plight of the more charismatic creatures—the whooping crane, the tiger, the elephant. Often the plethora of issues, political, social, economic and biological, involved in the effort to save a species, makes us throw up our hands in despair.

Our approach to conservation is usually a piecemeal one, a reaction to what we feel are the immediate compulsions of the moment. The elephant is under threat from poaching, so we send in the army to hunt down the poachers, or burn the ivory stocks. These are measures that are only necessary to the extent that they buy time. We fail to see the forest for the trees and the need for a more holistic approach to conservation. The behaviour of the Japanese in purchasing the ivory is as much responsible as the gun-toting ivory hunter for the decimation of elephants; the life-styles of the élite of New York, Nairobi and New Delhi contribute as much as the shifting cultivators do to the disappearance of the tropical rain forest. The battles to save the elephants and the rain forests have to be fought not only in Kenya, India, or Sumatra but also in Tokyo and Los Angeles.

I believe that the elephant (or the ozone layer, for that matter) can ultimately only be saved if we reduce our wasteful consumption of the planet's resources, lower the disparities between the rich and the poor, and work together in our stewardship of our beautiful earth.

We are today witnessing and abetting the most serious spasm of extinction of living creatures in the earth's history. During this century thousands, or perhaps hundreds of thousands of species, most of them undescribed, have vanished forever. An equal or greater number is already caught up in the extinction whirlpool and is destined to meet the same fate. Unless we act quickly and decisively the elephant will surely go the same way. And we ourselves will be reduced to a drab existence in a sterile concrete world without the exuberant laughter of a kookaburra floating over the clear waters or the magnificent trumpet of an elephant echoing through the forested hills.

The elephant has shared with us both war and peace, tribulation and joy, neglect and pomp. It is still a most evocative symbol, around which the conservation of Asia's rich natural heritage can be organized.

Sitting on the rocks along the Araikadavu I could just make out the silhouette of my tile-roofed dwelling against the southern skyline. It was dusk. Not a leaf seemed to stir, not a chirp from the ubiquitous cicada broke the silence. No spectacular streak of red, no gold-laced cloud adorned the western sky. The world was a canvas painted in shades of grey. There are rare moments in one's life when the mind seems strangely at peace with the world. This was one such moment. There were ten elephants a short distance from me. I knew no fear. Not at this time. The elephants were crossing the stream, their padded feet silently treading the rocks, careful not to violate the sanctity of the moment. One tiny elephant slipped on a wet rock, but its mother was behind, nudging it gently but powerfully with her trunk. Another elephant stopped to suck up water from a pool in the rocks. Soon the dark shapes merged with the fabric.

Elephant and human at peace with themselves and with the world, this is what I long for. I did not know this family, but at this moment, I was happy just to watch them as they were.

Postscript

My account of my decade-long study of elephants closed with the year 1990. This brief postscript takes it forward to September 1993.

The Veerappan saga continues with seemingly no end in sight. The bandit continues to roam the forests of Kollegal and nearby areas, in spite of a large force of both the police and army being sent after him. He continues to strike terror among the authorities, killing police officials at will. For me, the saddest part of this tragedy was the loss of another personal friend, P. Srinivas. I still remember the time when this idealistic young man came to Chamarajanagar as an Assistant Conservator in 1983, a few months before I left the place for Bangalore. His bubbling enthusiasm, deep-rooted sincerity and hard work soon earned him accolades not only within the forest department but also from the people of the Biligirirangans. When the Veerappan affair began to get out of hand Srinivas, who was then posted in another division, volunteered to shift his base to Gopinatham to bring the outlaw to justice. The idealistic young man that he was, he used methods unconventional in governmental circles in order to win the hearts of the villagers; he knew that Veerappan could not survive without the support of local villagers who provided him shelter, supplies and information. When this strategy seemed to be paying off in late 1991, Veerappan lured Srinivas into a trap, making the latter believe that he was ready to discuss the terms of a surrender, and shot him dead.

In spite of the ratio of adult male to female elephants remaining at about 1:10, the boom in births has been sustained among the elephants of Mudumalai and Bandipur. The interval between successive calvings has remained at about four to five years. Clearly, the loss of many big bulls to poachers has not yet been translated into fewer births as has been claimed for some African elephant populations. The remaining big bulls are doing their job of servicing the oestrous cows and the younger bulls are also reproductively active. Bulls which are only fifteen to twenty years old come into musth, even though this is not as intense as in older ones, and are certainly capable of mating.

This situation obviously cannot continue indefinitely. The loss of so many fine tuskers has certainly eroded the genetic base of the population. Even minor spurts in poaching, as during 1992 and early 1993, therefore exacerbate the situation.

Ironically, the elimination of the big bulls has also substantially reduced the incidence of crop depredation in the Eastern Ghats and the Nilgiris. There is still some raiding by family herds and the few remaining bulls, but this can in no way compare with the situation that prevailed in the early 1980s during the time of my study.

What impact will the expanding elephant population have on the vegetation? This will be clear only after several more years of study. In our fifty hectare plot at Mudumalai, *Kydia calycina* and *Helicteres isora* continued to decline rapidly until 1991, after which they seem to have stabilized.

Since 1991 our research at Mudumalai and the Nilgiris has diversified in several directions. We set up many smaller plots to study the interaction of elephants and vegetation over a much larger spatial scale, Milind Watve completed a study of parasites in the mammals, including elephants, at Mudumalai. His work threw up an intriguing possibility that elephants might 'spitefully' deposit their dung in water and infect other individuals visiting the pond after them! Justin Santosh joined our team to study the implications of the abnormally-skewed sex ratios on the development of behaviour in young male elephants. After completing a thesis on sociality in wasps, Arun Venkataraman turned his attention to a mammal with an equally fascinating social life—the pack-hunting and co-operatively breeding dholes at Mudumalai.

In 1991, a dynamic young officer, P. C. Tyagi, became the Warden of Mudumalai and initiated several measures to improve the management of this fine sanctuary, which was degrading as a result of steep increases in cattle numbers, the human population in nearby villages, developmental projects, and the influx of tourists. In addition, a new housing colony sprung up at Masinagudi on the shores of the Maravakandy reservoir in the wake of a new hydro-electric project, cutting off the view we had earlier enjoyed of elephants coming for a drink.

My contribution to the study and conservation of Asian elephants also brought unexpected honours. In March 1992 I received the Presidential Award of the Chicago Zoological Society from its Director, George Rabb, who is also the Chairman of IUCN's Species Survival Commission (SSC). The Asian Elephant Specialist Group, in the meanwhile, was making some progress in developing field projects and raising funds to sustain these. In this we have been supported by Simon Stuart, the head of SSC's office in Switzerland. I have been assisted by Uma Ramakrishnan in running the Asian Elephant Conservation Centre. Among the major sources of support have been the zoos of western countries. I have been invited to talk at several zoos and they have enormous potential for contributing to conservation efforts through research, on such aspects as genetics, which often cannot be tackled in wild populations, through imaginative breeding programmes of endangered species and by educating the public and in raising funds.

The efforts to save the Asian elephant now needs a massive thrust. Most countries are ill-equipped to implement a scientific conservation plan and need considerable international support. Indeed, the elephant continues to lose ground even in countries where expertise and funds are available. Driving through my study area in early 1993 I was made painfully aware of this slow yet steady process of attrition of the elephant's habitat. A crucial corridor between Punjur and Kolipalya used by Meenakshi's clan had been cleared for subsistence cultivation. Elephants need space, and lots of it. Unless humans can achieve a balance between their own legitimate needs and the needs of their fellow creatures, there may be little hope for the elephant.

As I complete this postscript, the seventh day after *Ganesh Chaturthi*, I hear the sharp burst of crackers and the joyful trumpet of music. I see the huge, colourful image of the elephant-headed god of wisdom, his benevolent face rising above an ecstatic crowd. Perhaps there is still hope for the elephant, a creature that has meant so much to so many people. Perhaps the wisdom of Ganesha will enlighten us as we approach the twenty-first century and give a new lease of life to the elephant, so that the magnificent trumpet of the elephant will then still echo through the forested hills of Asia a hundred years from now.

Glossary

bacha : a child (Hindi).
dosai : a flat pancake made of fermented rice and lentil paste (Tamil).
goonda : a thug (Hindi).
gopuram : the dome or tower of a temple usually featuring a variety of carvings (Tamil).
howdah : a seat for two or more, usually with a canopy for riding on the back of an elephant (Urdu).
idli : a fluffy pancake made of fermented rice and lentil paste (Tamil).
junglee : of the jungle or forest (Hindi).
kheddah : enclosure into which elephants are driven (Hindi).
koomeriah : thoroughbred or royal (Sanskrit).
koonkie : a tame elephant that assists in the capture and training of wild elephants (Hindi).
kumiri : shifting or swidden cultivation (Tamil/Kannada); *jhum* in Hindi.
kuthari : a stack of harvested cereal or millet plants kept in the fields for drying (Tamil).
machan : a tree-top platform or hut (Hindi).
makhna : a tuskless male elephant (Hindi).
mriga : a deer (Sanskrit).
patti : a cattle pen (Tamil/Kannada).
pil-khana : elephant stable (Hindi/Urdu).
podu : a small settlement in hill forests (Kannada).
ragi : finger millet *Eleusine coracana*.
saar : Sir (a south Indian peculiarity).
shola : refers to the distinctive patches of evergreen forest on mountain tops (from the Tamil *sholai*).
terai : moist land; refers to the region at the foothills of the Himalaya (Hindi).
vana : forest (Sanskrit).
vanavasi : a forest dweller (Sanskrit/Tamil).
virakal : hero stone (Tamil).
yanai : elephant (Tamil).

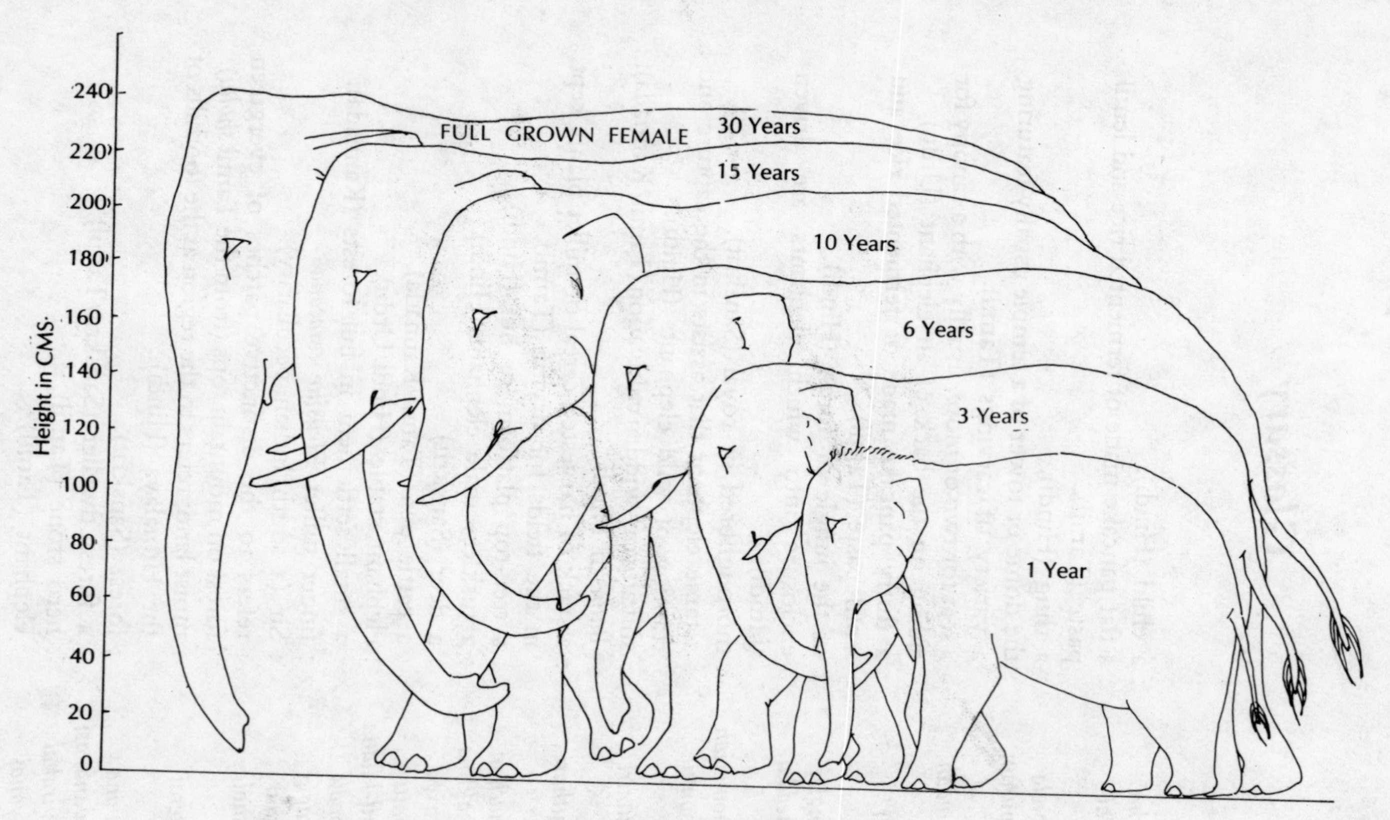

Field key for ageing young male elephants.

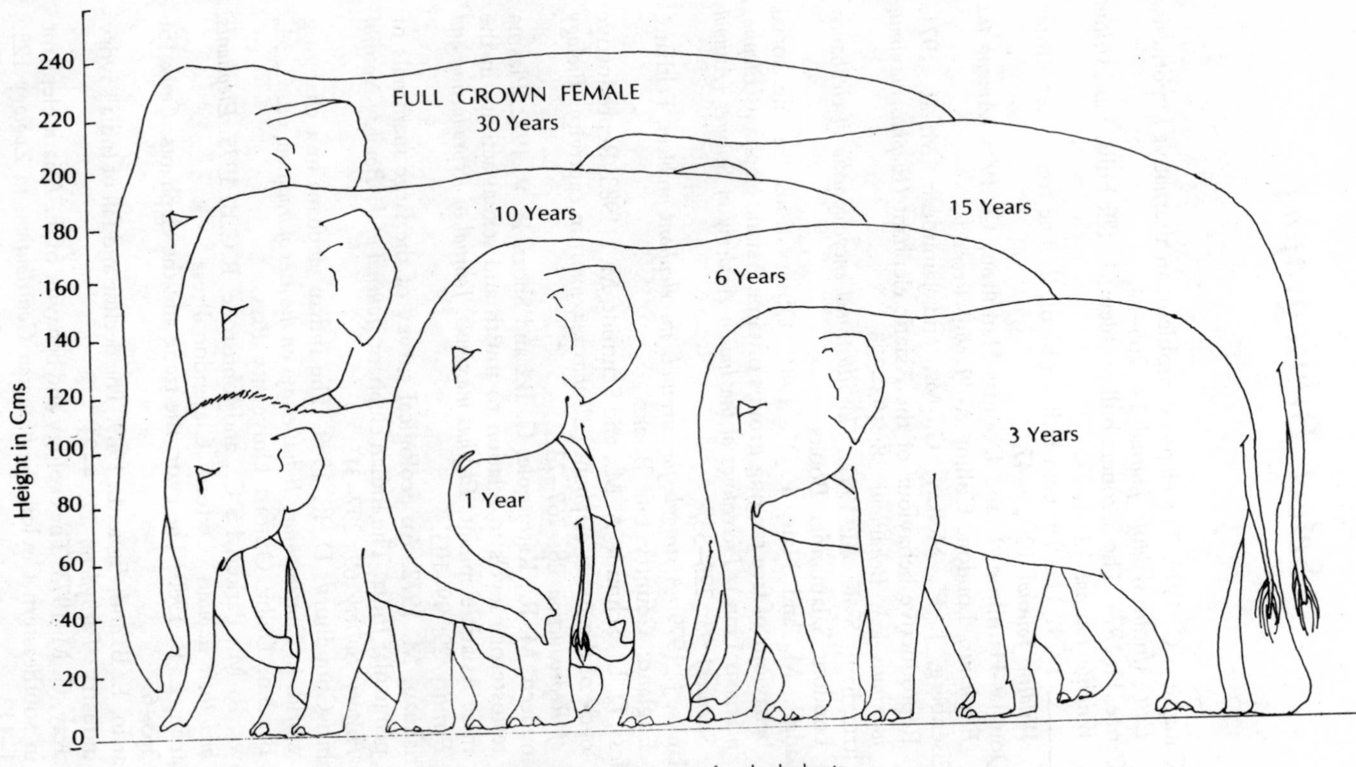

Field key for ageing young female elephants.

Select Bibliography

Caughley, G. 1976. The elephant problem: an alternative hypothesis. *East African Wildlife Journal* 14: 265–84.

Croze, H. 1974. The Seronera bull problem. 1. The bulls. *East African Wildlife Journal* 12: 1–27.

———, 1974. The Seronera bull problem. 2. The trees. *East African Wildlife Journal* 12: 29–47.

Douglas-Hamilton, I. and Douglas-Hamilton, O. 1975. *Among the Elephants*. London: Collins & Harvill Press.

Eisenberg, J. F., McKay, G. M., and Jainudeen, M. R. 1971. Reproductive behaviour of the Asiatic elephant (*Elephas maximus maximus* L.). *Behaviour* 38: 193–225.

Eltringham, S. K. ed. 1991. *The illustrated encyclopaedia of elephants*. London: Salamander Books.

Gadgil, M. and Nair, P. V. 1984. Observations on the social behaviour of free ranging groups of tame Asiatic elephant (*Elephas maximus* Linn.). *Proceedings of the Indian Academy of Sciences (Animal Sciences)* 93: 225–33.

Hanks, J. 1979. *A struggle for survival: the elephant problem*. Feltham, England: Country Life Books.

Hess, D. L., Schmidt, A. M., and Schmidt, M. J. 1983. Reproductive cycle of the Asian elephant (*Elephas maximus*) in captivity. *Biology of Reproduction* 28: 767–73.

Jainudeen, M. R., Katongole, C. B., and Short, R. V. 1972. Plasma testosterone levels in relation to musth and sexual activity in the male Asiatic elephant, *Elephas maximus*. *Journal of Reproduction and Fertility* 29: 99–103.

Krishnan, M. 1972. An ecological survey of the large mammals of peninsular India. The Indian elephant. *Journal of the Bombay Natural History Society* 69: 297–315.

Lahiri-Choudhury, D. K. 1989. The Indian elephant in a changing world. In *Contemporary India: essays on the uses of tradition*, ed. C. M. Borden. Delhi: Oxford University Press.

Laws, R. M., Parker, I.S.C., and Johnstone, R.C.B. 1975. *Elephants and their habitats*. Oxford: Clarendon Press.

Martin, E. B. 1980. The craft, the trade and the elephants. *Oryx* 15: 363–6.

Martin, E. B. and Vigne, L. 1989. The decline and fall of India's ivory industry. *Pachyderm* 12: 4–21.

McKay, G. M. 1973. The ecology and behavior of the Asiatic elephant in southeastern Ceylon. *Smithsonian Contributions to Zoology* 125: 1–113.

Moss, C. 1988. *Elephant memories: thirteen years in the life of an elephant family*. New York: William Morrow.

Nair, P. V. 1983. Studies on the development of behaviour in the Asiatic elephant. Ph.D. thesis, Indian Institute of Science, Bangalore. Unpublished.

Olivier, R. 1978. Distribution and status of the Asian elephant. *Oryx* 14: 379–424.

Owen-Smith, R. N. 1988. *Megaherbivores: the influence of very large body size on ecology*. Cambridge: Cambridge University Press.

Payne, K. B., Langbauer Jr., W. R., and Thomas, E. M. 1986. Infrasonic calls of the Asian elephant *(Elephas maximus)*. *Behavioural Ecology and Sociobiology* 18: 297–301.

Poole, J. H. and Moss, C. J. 1981. Musth in the African elephant, *Loxodonta africana*. *Nature* 292: 830–1.

Sanderson, G. P. 1878. *Thirteen years among the wild beasts of India*. London: W. H. Allen.

Santiapillai, C. and Jackson, P. 1990. The Asian elephant: an action plan for its conservation. Gland, Switzerland: IUCN–The World Conservation Union.

Sinclair, A. R. E. and Norton-Griffiths, M. eds. 1979. *Serengeti—dynamics of an ecosystem*. Chicago: University of Chicago Press.

Sukumar, R. 1989. *The Asian elephant: ecology and management*. Cambridge: Cambridge University Press.

______, 1989. Ecology of the Asian elephant in southern India. 1. Movement and habitat utilization patterns. *Journal of Tropical Ecology* 5: 1–18.

______, 1990. Ecology of the Asian elephant in southern India. 2. Feeding habits and crop raiding patterns. *Journal of Tropical Ecology* 6: 33–53.

______, 1991. The management of large mammals in relation to male strategies and conflict with people. *Biological Conservation* 55: 93–102.

Sukumar, R., Bhattacharya, S. K., and Krishnamurthy, R. V. 1987. Carbon isotopic evidence for different feeding patterns in an Asian elephant population. *Current Science* 56: 11–14.

Sukumar, R. and Gadgil, M. 1988. Male–female differences in foraging on crops by Asian elephants. *Animal Behaviour* 36: 1233–5.

Sukumar, R., Joshi, N. V., and Krishnamurthy, V. 1988. Growth in the Asian elephant. *Proceedings of the Indian Academy of Sciences (Animal Sciences)* 97: 561–71.

Sukumar, R. and Ramesh R. 1992. Stable carbon isotope ratios in Asian elephant collagen: implications for dietary studies. *Oecologia* 91: 536–9.

Western, D. and van Praet, C. 1973. Cyclical changes in the habitat and climate of an East African ecosystem. *Nature* 241: 104–6.

Index

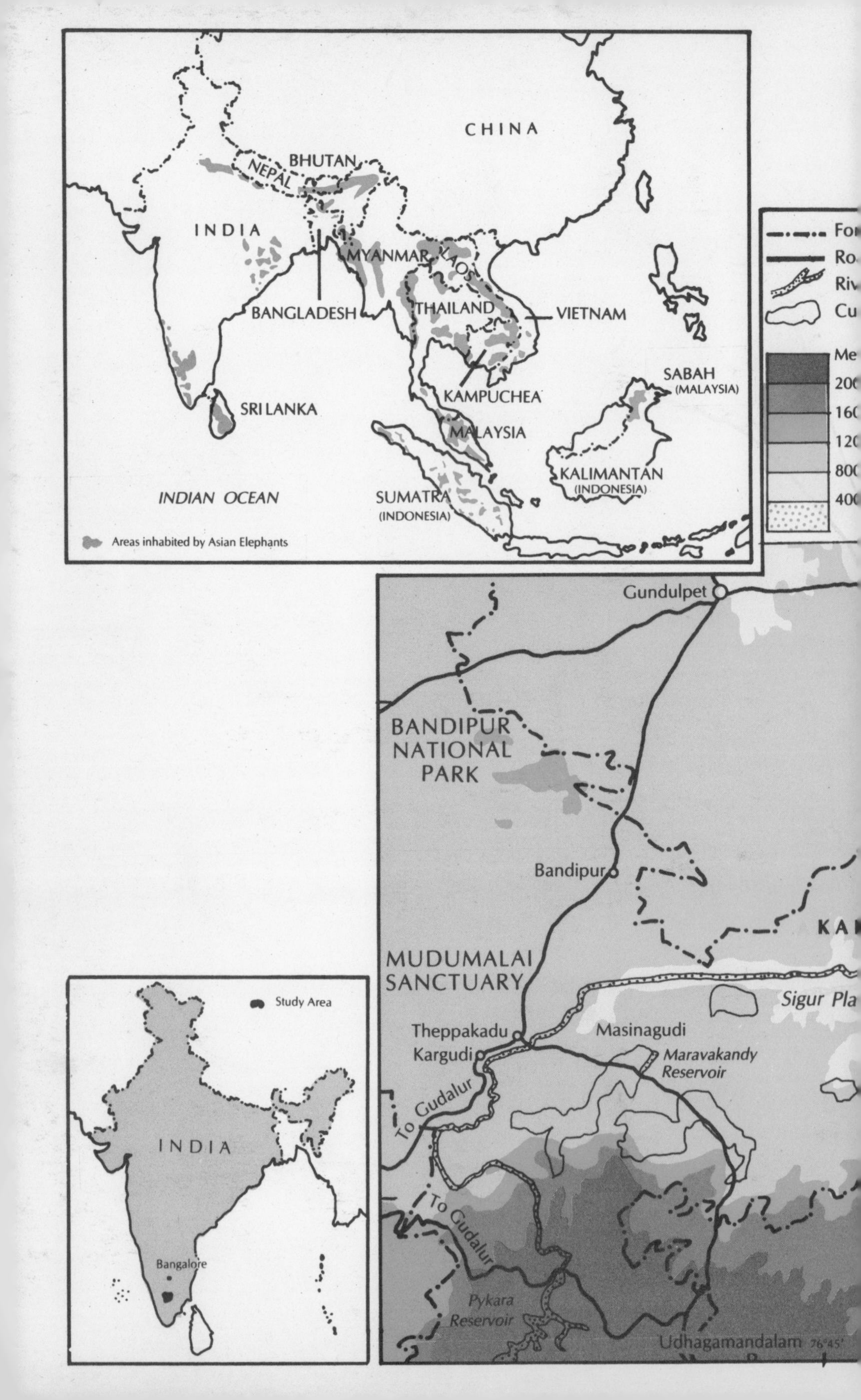
CHINA
BHUTAN
NEPAL
INDIA
MYANMAR
LAOS
BANGLADESH
THAILAND
VIETNAM
KAMPUCHEA
SABAH
(MALAYSIA)
SRI LANKA
MALAYSIA
KALIMANTAN
(INDONESIA)
INDIAN OCEAN
SUMATRA
(INDONESIA)
Areas inhabited by Asian Elephants
Gundulpet
BANDIPUR
NATIONAL
PARK
Bandipur
MUDUMALAI
SANCTUARY
Theppakadu
Kargudi
Masinagudi
Maravakandy
Reservoir
Sigur Pla
To Gudalur
To Gudalur
Pykara
Reservoir
Udhagamandalam
76°45'
Study Area
INDIA
Bangalore